Introduction to Potable Water Treatment Processes

Introduction to Potable Water Treatment Processes

**Simon A. Parsons
and Bruce Jefferson**
*School of Water Sciences
Cranfield University*

Blackwell
Publishing

Editorial Offices:
Blackwell Publishing Ltd, 9600 Garsington Road, Oxford OX4 2DQ, UK
  Tel: +44 (0) 1865 776868
Blackwell Publishing Professional, 2121 State Avenue, Ames, Iowa 50014-8300, USA
  Tel: +1 515 292 0140
Blackwell Publishing Asia Pty Ltd, 550 Swanston Street, Carlton, Victoria 3053, Australia
  Tel: +61 (0)3 8359 1011

First published 2006 by Blackwell Publishing Ltd

Library of Congress Cataloging-in-Publication Data

Parsons, Simon, Dr.
  Introduction to potable water treatment processes / Simon A. Parsons and Bruce Jefferson.
     p. cm.
  Includes bibliographical references and index.
    ISBN-13: 978-1-4051-2796-7 (alk. paper)
    ISBN-10: 1-4051-2796-1 (alk. paper)
  1. Water–Purification.   2. Drinking water–Purification.   I. Jefferson, Bruce. II. Title.

  TD430.P37   2006
  628.1'62–dc22
                                        2005021092

ISBN-10: 1-4051-2796-1
ISBN-13: 978-1-4051-2796-7

A catalogue record for this title is available from the British Library

Set in 10.5/13.5pt Sabon and Optima by TechBooks, New Delhi, India
Printed and bound in India by Replika Press Pvt Ltd

For further information on Blackwell Publishing, visit our website:
www.blackwellpublishing.com

Contents

Preface

Being able to turn on a tap and fill a glass of water to drink is something we tend to take for granted but the quality of the drinking water at any tap depends on many factors going all the way back to the original source of the water, how it has been treated and how it has been distributed to the tap. The quality of water we receive in the UK today is achieved through a source to tap approach including continual improvements in source protection, water treatment, operation and maintenance, quality monitoring as well as training and education.

The material in this book was originally produced for a short course called Introduction to Water Treatment Processes. The course was designed to give an overview to water treatment processes and to try and show how to link source water quality to treatment process selection and it was organised as part of the activities of the Network on Potable Water Treatment and Supply. The Network is funded by the Engineering and Physical Sciences Research Council (EPSRC) and aims to encourage multidisciplinary research, education and development in the areas of potable water treatment and supply.

In writing the book we have kept a number of audiences in mind, particularly undergraduate student studying for civil engineering, environmental engineering and environmental science degrees and postgraduate students studying for masters and doctorates in the field of water and wastewater treatment. It is also aimed at scientist and engineers working in the water industry who might be interested in gaining a broader knowledge of the process involved in water treatment.

We have many people to thanks for the providing us with much of the data, images and information presented in the book. In no apparent order these include Jenny Banks, Andy Campbell, Ed Irvine, Tim Pearson, Alasdair Maclean, Barrie Holden, David Essex, Peter Hillis, Michael

Chipps, Melisa Steele, Jeffrey Nelson, Terry Crocker, Jason Lerwill, Elise Cartmell, Brian McIntosh, Derek Wilson, Paul Jeffrey, Richard Stuetz, Simon Judd, Peter Daniels, Paul Godbold, Tom Hall, Rita Henderson, Philippe Sauvignet, Emma Sharp, Tony Amato, Stuart Watson, Kelvin O'Halloran and Jonathan Waites.

We would also like to acknowledge the patience of Blackwell Publishing staff involved in this project especially Sarahjayne and Pooja.

Simon Parsons and Bruce Jefferson
December 2005

Water Quality Regulations 1

1.1 Introduction

The availability of a reliable and clean supply of water is one of the most important determinants of our health. Historically, improvements in human health have been related to improvements in our water supply system from source to tap. The quality of water we receive in the UK today is achieved through ongoing improvements in source protection, water treatment, operation and maintenance, quality monitoring and training and education. This chapter will focus on the current UK and WHO regulations for potable water and identify some of the key contaminants that affect treatment process selection and operation.

1.2 Water quality regulations

1.2.1 Europe/UK

The European Directive (Council Directive 98/83/EC, 1998) on the quality of water intended for human consumption prescribes standards for the quality of drinking water, water offered for sale in bottles or containers and water for use in food production undertakings. Its requirements have been incorporated into the Water Supply (Water Quality) Regulations 2000 in England and Wales. The Directive specifies two types of parameter values, namely mandatory and non-mandatory. Mandatory standards, covering 28 microbiological and chemical parameters for mains water, are essential for health and the environment, and have to be met by specified dates. Any contravention of an indicator value must be investigated, but remedial action need be taken only where there is a risk to public health. These are given in Tables 1.1 and 1.2 whilst indicator and radioactivity parameters are given in Tables 1.3 and 1.4, respectively. In addition, there

Table 1.1 Microbiological parameters (Council Directive 98/83/EC).

Parameter	Parametric value	Unit
Escherichia coli (E. coli)	0	number/100 ml
Enterococci	0	number/100 ml

are non-mandatory indicator values, covering 20 further microbiological, chemical and physical parameters that are prescribed for monitoring purposes. These include parameters such as turbidity, which are important for avoiding discoloured water problems, other aesthetic parameters such as taste, odour and colour, and also the pH value which is important in relation to lead (plumbosolvency) and copper (cuprosolvency). In addition the UK has specific regulations in place for *Cryptosporidium* where a treatment standard has been set of, on average, less than one oocyst in 10 l of water supplied from a treatment works (Water Supply (Water Quality) Regulations 1989, 2000b). The standard does not take into account different species of *Cryptosporidium*, nor whether any oocysts detected are viable, i.e. alive and potentially able to cause infection.

Table 1.2 Chemical parameters (Source: DWI).

Parameter	Parametric value	Unit
Acrylamide	0.10	$\mu g\ l^{-1}$
Antimony	5.0	$\mu g\ l^{-1}$
Arsenic	10	$\mu g\ l^{-1}$
Benzene	1.0	$\mu g\ l^{-1}$
Benzo(a)pyrene	0.010	$\mu g\ l^{-1}$
Boron	1000	$\mu g\ l^{-1}$
Bromate	10	$\mu g\ l^{-1}$
Cadmium	5.0	$\mu g\ l^{-1}$
Chromium	50	$\mu g\ l^{-1}$
Copper	2000	$\mu g\ l^{-1}$
Cyanide	50	$\mu g\ l^{-1}$
1,2-Dichloroethane	3.0	$\mu g\ l^{-1}$
Epichlorohydrin	0.10	$\mu g\ l^{-1}$
Fluoride	1500	$\mu g\ l^{-1}$
Lead	10	$\mu g\ l^{-1}$
Mercury	1.0	$\mu g\ l^{-1}$
Nickel	20	$\mu g\ l^{-1}$
Nitrate	50 000	$\mu g\ l^{-1}$
Nitrite	500	$\mu g\ l^{-1}$
Pesticides	0.10	$\mu g\ l^{-1}$
Pesticides – Total	0.50	$\mu g\ l^{-1}$
Polycyclic aromatic hydrocarbons	0.10	$\mu g\ l^{-1}$
Selenium	10	$\mu g\ l^{-1}$
Tetrachloroethene and Trichloroethene	10	$\mu g\ l^{-1}$
Trihalomethanes – Total	100	$\mu g\ l^{-1}$
Vinyl chloride	0.50	$\mu g\ l^{-1}$

Table 1.3 Indicator parameters (Council Directive 98/83/EC).

Parameter	Parametric value	Unit
Aluminium	200	$\mu g \ l^{-1}$
Ammonium	500	$\mu g \ l^{-1}$
Chloride	250	$mg \ l^{-1}$
Clostridium perfringens (including spores)	0	number/100 ml
Colour	Acceptable to consumers and no abnormal change	
Conductivity	2500	$\mu S \ cm^{-1}$ at 20°C
Hydrogen ion concentration	≥ 6.5 and ≤ 9.5	pH units
Iron	200	$\mu g \ l^{-1}$
Manganese	50	$\mu g \ l^{-1}$
Odour	Acceptable to consumers and no abnormal change	
Oxidisability	5.0	$mg \ l^{-1} \ O_2$
Sulphate	250	$mg \ l^{-1}$
Sodium	200	$mg \ l^{-1}$
Taste	Acceptable to consumers and no abnormal change	
Colony count	No abnormal change	
Coliform bacteria	0	number/100 ml
Total organic carbon (TOC)	No abnormal change	
Turbidity	Acceptable to consumers and no abnormal change	

World Health Organisation

An established goal of the World Health Organisation (WHO) is that all people, whatever their stage of development and their social and economic conditions, have access to an adequate supply of safe drinking water (WHO, 2004). To achieve this WHO has been involved in the review and evaluation of information on health aspects of drinking-water supply and quality and in issuing guidance material on the subject. The first WHO publication on drinking water quality was published in 1958 as International Standards for Drinking Water and was revised in 1963 and in 1971 under the same title. In 1984 the WHO *Guidelines for Drinking-Water Quality* were published and revised in 1997 and again in 2004. The guidelines propose values for microbiological and chemical species (see Table 1.5). For example guideline values were set for the bacteria *E. coli* or thermotolerant (faecal) coliforms and total coliforms of none detectable

Table 1.4 Radioactivity (Council Directive 98/83/EC).

Parameter	Parametric value	Unit
Tritium	100	Becquerel/l
Total indicative dose	0.10	mSv/year

Table 1.5 Guideline values for chemicals that are of health significance in drinking water (Source: WHO, 2004).

Guideline value[a]	Chemical (μg l^{-1})	Remarks
Acrylamide	0.5b	
Alachlor	20b	
Aldicarb	10	Applies to aldicarb sulfoxide and aldicarb sulfone
Aldrin and dieldrin	0.03	For combined aldrin plus dieldrin
Antimony	20	
Arsenic	10 (P)	
Atrazine	2	
Barium	700	
Benzene	10b	
Benzo[a]pyrene	0.7b	
Boron	500 (T)	
Bromate	10b (A, T)	
Bromodichloromethane	6b	
Bromoform	100	
Cadmium	3	
Carbofuran	7	
Carbon tetrachloride	4	
Chloral hydrate (trichloroacetaldehyde)	10 (P)	
Chlorate	700 (D)	
Chlordane	0.2	
Chlorine	5 (C)	For effective disinfection, there should be a residual concentration of free chlorine of $\geq$0.5 mg l^{-1} after at least 30 min contact time at pH <8.0
Chlorite	700 (D)	
Chloroform	200	
Chlorotoluron	30	
Chlorpyrifos	30	
Chromium	50 (P)	For total chromium
Copper	2000	Staining of laundry and sanitary ware may occur below guideline value
Cyanazine	0.6	
Cyanide	70	
Cyanogen chloride	70	For cyanide as total cyanogenic compounds
2,4-D (2,4-dichlorophenoxyacetic)	30	Applies to free acid acid
2,4-DB	90	
DDT and metabolites	1	
Di(2-ethylhexyl)phthalate	8	
Dibromoacetonitrile	70	
Dibromochloromethane	100	
1,2-Dibromo-3-chloropropane	10b	
1,2-Dibromoethane	0.4b (P)	
Dichloroacetate	50 (T, D)	
Dichloroacetonitrile	20 (P)	
Dichlorobenzene 1,2-	1000 (C)	

Table 1.5 (*Continued*)

Guideline value[a]	Chemical ($\mu g\ l^{-1}$)	Remarks
Dichlorobenzene, 1,4-	300 (C)	
Dichloroethane, 1,2-	30b	
Dichloroethene, 1,1-	30	
Dichloroethene, 1,2-	50	
Dichloromethane	20	
1,2-Dichloropropane (1,2-DCP)	40 (P)	
1,3-Dichloropropene	20b	
Dichlorprop	100	
Dimethoate	6	
Edetic acid (EDTA)	600	Applies to the free acid
Endrin	0.6	
Epichlorohydrin	0.4 (P)	
Ethylbenzene	300 (C)	
Fenoprop	9	
Fluoride	1500	Volume of water consumed and intake from other sources should be considered when setting national standards
Formaldehyde	900	
Hexachlorobutadiene	0.6	
Isoproturon	9	
Lead	10	
Lindane	2	
Manganese	400 (C)	
MCPA	2	
Mecoprop	10	
Mercury	1	For total mercury (inorganic plus organic)
Methoxychlor	20	
Metolachlor	10	
Microcystin-LR	1 (P)	For total microcystin-LR (free plus cellbound)
Molinate	6	
Molybdenum	70	
Monochloramine	3000	
Monochloroacetate	20	
Nickel	20 (P)	
Nitrate (as NO_3^-)	50 000	Short-term exposure
Nitrilotriacetic acid (NTA)	200	
Nitrite (as NO_2^-)	3000	Short-term exposure
	200 (P)	Long-term exposure
Pendimethalin	20	
Pentachlorophenol	9b (P)	
Pyriproxyfen	300	
Selenium	10	
Simazine	2	
Styrene	20 (C)	
2,4,5-T	9	
Terbuthylazine	7	
Tetrachloroethene	40	
Toluene	700 (C)	

(*Continued*)

Table 1.5 Guideline values for chemicals that are of health significance in drinking water (Source: WHO, 2004). (*Continued*)

Guideline value[a]	Chemical (μg l^{-1})	Remarks
Trichloroacetate	200	
Trichloroethene	70 (P)	
Trichlorophenol, 2,4,6-	200b (C)	
Trifluralin	20	
Trihalomethanes		The sum of the ratio of the concentration of each to its respective guideline value should not exceed 1
Uranium	15 (P, T)	Only chemical aspects of uranium addressed
Vinyl chloride	0.3b	
Xylenes	500 (C)	

in any 100 ml sample. These bacteria were selected as they give a good indication of the likelihood of faecal contamination and the integrity of a water supply. Details of chemical guideline values are shown below.

P = provisional guideline value, as there is evidence of a hazard, but the available information on health effects is limited; T = provisional guideline value because calculated guideline value is below the level that can be achieved through practical treatment methods, source protection, etc.; A = provisional guideline value because calculated guideline value is below the achievable quantification level; D = provisional guideline value because disinfection is likely to result in the guideline value being exceeded; C = concentrations of the substance at or below the health based guideline value may affect the appearance, taste or odour of the water, leading to consumer complaints; b = for substances that are considered to be carcinogenic, the guideline value is the concentration in drinking water associated with an upper-bound excess lifetime cancer risk of 10^{-5} (one additional cancer per 100 000 of the population ingesting drinking water containing the substance at the guideline value for 70 years).

1.3 Common contaminants

1.3.1 Turbidity

Turbidity is a general measure of water 'cloudiness' created by particles suspended in a water sample. It has been used to assess drinking water quality for a century and is still the most widely used particle measurement in water treatment. These particles may include clay, silt, finely divided inorganic and organic matter, soluble coloured organic compounds, and

Table 1.6 Outbreaks of illness associated with public and private drinking water supplies in the UK 1991–2000 (adapted from Standing Committee of Analysts, 2002).

Pathogen	Public supplies	Private supplies
Cryptosporidium	24	4
Giardia		1
Campylobacter	4	16
E. coli O157	2	4
Salmonella	1	
Unknown	2	

plankton and other microscopic organisms. Excessive turbidity, or cloudiness, in drinking water is aesthetically unappealing, and may also represent a health concern as it may provide food and shelter for pathogens.

1.3.2 Microbiology

The largest public health impact of unsafe drinking water is diarrhoeal disease and the majority of water-associated outbreaks of disease can be related back to the microbiological quality of drinking water. These include infectious and parasitic diseases such as cholera, typhoid, dysentery, hepatitis, giardiasis, guinea worm and schistosomiasis. Specific bacteria of concern include *Campylobacter*, *E. coli* O157, *Salmonella*, all of which have been linked to outbreaks of illness from consuming drinking water (Table 1.6). Most strains of bacteria are harmless such as *E. coli* which live in the intestines of healthy animals and humans but one species *E. coli* O157:H7 is a strain that produces a toxin that can cause severe illness. In addition there are a number of viruses of concern including Norwalk-like viruses (or small round structured viruses), which are the most common cause of sporadic and epidemic viral gastroenteritis in adults (Standing Committee of Analysts, 2002). Other viruses include the Hepatitis A virus, Enteroviruses, Rotaviruses and Adenoviruses. The most important protozoan pathogens are *Cryptosporidium* and *Giardia* with the majority of recent outbreaks of illness associated with drinking water being due to *Cryptosporidium*. Table 1.7 compares the concentrations of bacterial,

Table 1.7 Examples of detectable concentrations (per litre) of enteric pathogens and faecal indicators in different types of source waters (adapted from WHO, 2004).

Pathogen	Lakes and reservoirs	Impacted rivers and streams	Groundwater
E. coli (generic)	10 000–1 000 000	30 000–1 000 000	0–1000
Viruses	1–10	30–60	0–2
Cryptosporidium	4–290	2–480	0–1
Giardia	2–30	1–470	0–1

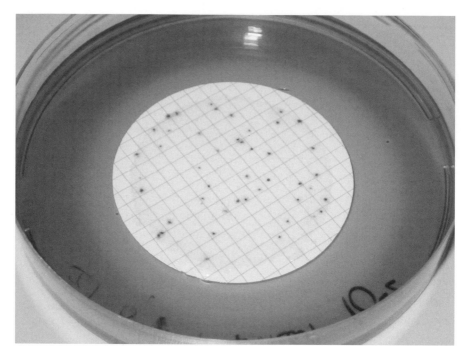

Fig. 1.1   Plate showing coliforms and *E. coli* growing on membrane lactose glucuronide agar (MLGA) a selective chromogenic medium for differentiation of *E. coli* and other coliforms. Coliforms are light colonies and *E. coli* are the dark colonies.

viruses and protozoa found in different water sources. Figure 1.1 shows the results of a typical test to quantify the concentration of Coliforms and *E. coli* found in a water sample.

Cryptosporidium

Cryptosporidium is a protozoan that can cause an acute form of gastroenteritis, known as cryptosporidiosis, in humans and animals. In the last two decades, concern in the water industry has grown over the specific human pathogen, *Cryptosporidium parvum*, which exhibits some resistance to traditional disinfection methods. The first diagnosed case of cryptosporidiosis in humans was reported in 1976 and a number of high profile, waterborne outbreaks of the disease have been reported. Contaminated groundwater was suspected to be the cause of the first documented waterborne outbreak of cryptosporidiosis at Braun Station, Texas in 1984. In 1993, *Cryptosporidium* caused 400 000 people in Milwaukee to experience intestinal illness, 4000 were hospitalised and at least 50 deaths have been attributed to the disease. Similar incidents in the UK have included the 1997 Three Valleys outbreak, which affected 345 people in North London and Hertfordshire.

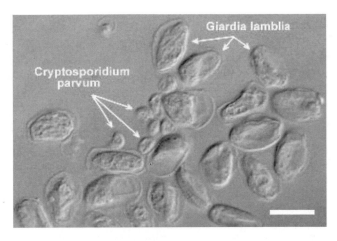

Fig. 1.2 *Cryptosporidium parvum* oocysts and *Giardia lamblia* (intestinalis) cysts; the *Cryptosporidium parvum* oocysts are distinguished from neighbouring *G. lamblia* cysts by their smaller size. Bar = 10 μm (Source: H.D.A. Lindquist, USEPA).

Cryptosporidiosis is spread by the faecal–oral route. Outside of an infected host, the parasite exists as a hard-coated, dormant form called an 'oocyst'. These are spherical, some 4–6 μm in diameter (Figure 1.2). If ingested in sufficient number, the organism is able to infect the small intestine where it undergoes a complex lifecycle. The final stage sees an infected host typically passing millions of *Cryptosporidium* 'oocysts' in their faeces, often over a period of several weeks after initial infection. The disease is usually spread by direct contact with an infected person or animal. However, it is a particular concern in the water industry because oocysts can contaminate raw water supplies via the discharge or run-off into rivers of human and animal wastes. Indeed, they are found in low concentrations in many raw waters (Table 1.7).

Algae
Algae are photosynthetic organisms, which range in size from 2 μm to those large enough to be seen by the naked eye. They thrive in an aquatic or moist environment with a high nutrient availability and they are ubiquitous in surface waters at low concentrations. However, many surface waters suffer eutrophic conditions in that either an influx of fertiliser or sewage effluent significantly increases the supply of nutrients such that excessive algal growth occurs. Algal populations can reach concentrations in excess of 10^6 cells ml^{-1}, which can significantly interfere with water treatment processes. In the past decade there have been over 4000 reported incidences of blooms affecting over 2000 UK water bodies, with estimated costs of between £75–114 million a year to remedy (Pretty *et al.*, 2003). A

substantial proportion of this cost comes from its impact on drinking water treatment and the need for additional process stages. Algae can cause many problems including physical blocking of treatment processes, release of taste and odour causing compounds and some species are known to produce toxins.

There are many different species of algae, and common British species of fresh water algae can be split among the following eight divisions: *cyanophyta* (blue-green algae); *chlorophyta* (green algae); *xanthophyta* (yellow-green algae); *chrysophyta* (golden algae); *bacillariophyta* (diatoms); *pyrrophyta*; *cryptophyta* (cryptomonads) and *euglenophyta* (euglenoids). However, during periods of eutrophication, diatoms, green algae and blue-green algae tend to dominate source water in spring, early summer and late summer/autumn, respectively.

Toxic algal metabolites are typically associated with cyanobacteria blooms, as opposed to diatoms or green algae, where it has been reported that approximately 60% of cyanobacteria blooms are toxic. In high concentrations the so-called cyanotoxins produced pose a serious risk to human health and hence, must be removed from the water. Cyanotoxins are divided into hepatotoxins, such as microcystin-LR (MCLR) which induce liver failure; neurotoxins, which affect the nervous system; and cytotoxins, are irritant and allergenic substances. There have been recorded incidences of illness and deaths due to the presence of a variety of cyanobacterium in water resulting in the WHO setting a guideline value of $1\ \mu g\,l^{-1}$ for MCLR.

Pesticides

Pesticides play an important role in modern agriculture and typically comprise a formulation containing the active ingredient and co-formulants. In the UK, the greatest amounts of pesticides are used in arable farming and the treatment of grassland and fodder (Environment Agency of England and Wales, 2002, Figure 1.3). In terms of weight of substances

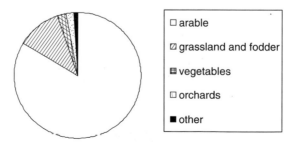

Fig. 1.3   Relative areas of crops treated with pesticides annually in the UK (adapted from Thomas & Wardman, 1999).

applied to arable land in the UK in 2002, herbicides and desiccants were used in the largest amounts (71%), followed by fungicides (12%), growth regulators (11%), insecticides and nematicides (2%), molluscicides (1%) and seed treatments (1%). The most extensively used herbicides were glyphosate, isoproturon, fluroxypyr, mecoprop-P and diflufenicam. The most used fungicides were epoxiconazole, azoxystrobin, tebuconazole, kresoxim methyl, fenpropimorph and trifloxystrobin and the most widely used insecticides were the pyrethroids (cypermethrin, esfenvalerate), organophosphates and the carbamates (pirimicarb).

Recent US and European monitoring data has shown detectable concentrations of pesticides in more than 95% of sampled surface waters and approximately 50% of sampled ground waters, making them of concern to the drinking water industry (Parsons *et al.*, 2005)

Lead

Lead (Pb) is a metal found in natural deposits and in household plumbing materials and water pipes. The greatest exposure to lead is swallowing or breathing in lead paint chips and dust, and lead is rarely found in source waters. It is also known to enter drinking water through the corrosion of plumbing materials. Lead can cause a variety of adverse health effects. In babies and children, exposure to lead in drinking water above the action level can result in delays in physical and mental development, along with slight deficits in attention span and learning abilities. In adults, it can cause increases in blood pressure and if consumed over many years can lead to kidney problems. UK regulations for lead are 25 μg Pb l^{-1} from December 2003 until December 2013 when it will be reduced to 10 μg Pb l^{-1} at the consumers' taps.

Nitrate

Nitrate (NO_3^-) is increasingly polluting ground and surface waters primarily as a result of diffuse pollution from agricultural fertilisers. Nitrate is an oxidised inorganic contaminant which is soluble in water and reports of health problems associated with nitrate appeared in the 1940s where infants who had ingested high nitrate water had developed methaemoglobinaemia ('blue-baby syndrome'). The maximum admissible concentration for nitrate in drinking water is 11.3 mg NO_3-N l^{-1}.

Manganese

Manganese (Mn) is a naturally occurring element found in rocks and soils. It is released into flowing water naturally though contact with rocks and glacial sediment under acidic conditions. Although high concentrations due to industrial and agricultural sources have been reported, general naturally occurring manganese concentrations are below 0.5 mg l^{-1} even in acidic environments. In domestic water supplies manganese is seen as an undesirable contaminant more from an aesthetic and an economic viewpoint rather than a health risk (toxic at concentrations greater than 10 mg l^{-1}). Particular problems noted for manganese in the literature include manganese affects taste, colour and odour; manganese oxide precipitates stain materials and, where accumulated, clog treatment systems and pipe work.

NOM and disinfection by-products

Most water sources throughout the world contain natural organic matter (NOM), and it is best described as a complex mixture of organic compounds. NOM itself is considered harmless but its removal has become increasingly important in light of the potential for forming carcinogenic disinfection by-products (DBP) if organic carbon is insufficiently removed. As a result legislation around the world calls for the strict control of DBPs. The disinfection of drinking water uses powerful chemical disinfectants such as chlorine, chloramines, chlorine dioxide and ozone. These chemicals are also strong oxidants that convert organic matter in water into organic chemical DBPs. To date over 500 different DBPs have been identified. The major DBPs formed by chlorination are the trihalomethanes and haloacetic acids but over 60% of the compounds formed are unknown (Richardson, 2002). Fifty high-priority DBPs have been listed including iodo-THMs, bromo acids, bromoacetonitriles, bromoaldehydes, 3-chloro-(4-dichloromethyl)-5-hydroxy-2(5H)-furanone (MX), brominated forms of MX (the so-called BMXs) and other nonhalogenated compounds such as cyanoformaldehyde.

Bromate

Bromate (BrO$_3$$^-$), an oxyanion of bromine, is generally known as a disinfection by-product formed during ozonation of drinking water supplies containing bromide (Br$^-$). Bromate has been classed by the WHO as a 'possible human carcinogen', which has subsequently led to implementation of 10–25 μg l^{-1} drinking water limits in areas including the United States, European Union and United Kingdom. In contrast to bromide, bromate is not reported as occurring naturally in surface waters or aquifers. However, following recent advances in analytical techniques bromate has been detected in the surface water environment, possibly as a result of industrial oxidation/disinfection processes. In addition, bromate contamination has now been detected within an aquifer in the UK as a result of industrial contamination.

Pharmaceuticals

Pharmaceutical compounds (e.g. antibiotics), endocrine disrupting compounds (EDCs) and personal care products (e.g. shampoos, deodorants) can pollute water sources. Sewerage systems are the main collection point for pharmaceuticals. Since pharmaceuticals are mostly polar and potentially resistant to biotransformation, some are not degraded or removed during wastewater treatment, and may therefore, reach the wider water environment. Once discharged into receiving waters, many pharmaceuticals and their metabolites have been shown to persist and bio-concentrate, causing problems such as the feminisation of male fish and increased levels of antibiotic resistance.

Arsenic

Arsenic (As) is a silver-grey brittle crystalline solid and is found in many rock-forming minerals. It can also be present in many water sources, typically where water has flowed through arsenic-rich rocks (see Table 1.8). If ingested arsenic can lead to skin disorders including hyper/hypopigmentation changes and keratosis and skin cancer. The characteristic chemical features of the high-arsenic ground waters are high iron, manganese, bicarbonate and often phosphorus concentrations. Arsenic in drinking water has recently gained much attention due to problems in countries such as Bangladesh. Ninety per cent of the Bangladesh population of 130 million drink well water as an alternative to polluted surface water and thereby reducing the incidence of waterborne diseases. This has lead to between 8–12 million shallow tube wells being dug in Bangladesh where well waters have arsenic concentrations between $<0.5 \mu g \, l^{-1}$ and $3200 \mu g \, l^{-1}$.

Table 1.8 Examples of arsenic concentrations found in water sources (adapted from Smedley & Kinniburgh, 2005).

Source	Concentration ($\mu g \, l^{-1}$)
Rainfall (USA)	0.013–0.032
River water	0.13–2.1
Estuary water (Tamar estuary, UK)	2.7–8.8
Ground water (UK)	<0.5–10
Ground water (As-rich provinces: Bengal basin, Argentina, Mexico, northern China, Taiwan, Hungary)	10–5000

References

Council Directive 98/83/EC of 3 November 1998 on the quality of water intended for human consumption. *Off J Eur Commun*, 5.12.98, L330/32–L330/53.

Environment Agency of England and Wales (2002) Pesticides 2000: A summary of monitoring of the aquatic environment in England and Wales. Bristol, UK.

Parsons, S.A., Boxall, A.B.A & Sinclair, C. (2005) Pesticides, their degradates, and adjuvants of concern to the drinking water community. AwwaRF, Denver.

Pretty, J.N., Mason, C.F., Nedwell, D.B., Hine, R.E., Leaf, S. & Dils, R. (2003) Environmental costs of freshwater eutrophication in England and Wales. *Environ. Sci. Technol.*, 32, 201–208.

Richardson, S.D. (2002). Environmental mass spectrometry: emerging contaminants and current issues. *Anal. Chem.*, 74, 2719–2742.

Smedley, P.L. & Kinniburgh, D.G. (2005) Source and behaviour of arsenic in natural waters. United Nations Synthesis Report on Arsenic in Drinking Water. WHO, Geneva.

Standing Committee of Analysts (2002) The Microbiology of Water: Part 1 – Drinking Water, Methods for the Examination of Waters, Environment Agency.

Thomas, M.R. & Wardman, O.L. (1999) Review of usage of pesticides in agriculture and horticulture throughout Great Britain. MAFF, London.

The Water Supply (Water Quality) (Amendment) Regulations 1989, The Stationery Office Limited, London, ISBN 0110973844.

The Water Supply (Water Quality) Regulations 2000 (2000) The Stationery Office Limited, London, ISBN 0 11 090431.

WHO (2004) Guidelines for Drinking-Water Quality. World Health Organization, Geneva.

Water Sources and Demand 2

2.1 Introduction

Water is in a constant state of change and the conversion of water from atmospheric moisture to land-based water and back again is known as the water cycle. The notion that water is continually circulating from the ocean to the atmosphere to the land and back again to the ocean has interested scholars through most of recorded history. In Book 21 of the Iliad, Homer (ca. 810 BC) wrote of 'the deep-flowing Oceanus, from which flow all rivers and every sea and all springs and deep wells.' Thales (ca. 640–546 BC) and Plato (ca. 427–347 BC) also alluded to the water cycle when they wrote that all waters returned by various routes to the sea. However, it was not until many centuries later that scientific measurements confirmed the existence of a water (or hydrologic) cycle. Seventeenth century French physicists Pierre Perrault (1608–1680) and Edmond Mariotte (1620–1684) separately made crude precipitation measurements in the Seine River basin in France and correlated these measurements with the discharge of the river to demonstrate that quantities of rainfall and snow were adequate to support the river's flow.

The natural water cycle has been altered by man's activities such as the impounding of large quantities of fresh water, and there is now a man made water cycle that includes potable supply, collection and disposal of wastewaters. Fresh water is taken, or abstracted, from the environment for our use. In the UK the water industry provides 18 000 million litres of water every day to 58 million people and to achieve this, water is abstracted from a wide range of sources of varying water quality. Water sources are a combination of surface and groundwater and this abstracted water is treated at over 2200 water treatment works. The treatment process operated at each of these works is driven by the quality of the incoming

Table 2.1 Sources of water.

Water source	%Total
Salt water	97
Fresh water	
Icecaps	2.4
Ground water	0.58
Rivers and lakes	0.02
Atmosphere	<0.001
Total water	100

source waters as well as the requirements of the 400 000 km of water mains supplying the water to our taps.

This chapter will cover the water cycle including looking at water demand and will show how water quality is linked to treatment process selection.

2.2 Water cycle

So, how much water is there on the planet and where is it? Well, most of it is saline (i.e. salty), with estimates ranging from 94–97% of the total (Table 2.1). Much of the remaining freshwater is also unavailable as it is locked away in the ice caps and glaciers. Non-saline water that can be readily abstracted for domestic and industrial use is only available from lakes, rivers and from aquifers.

In England and Wales the majority of abstracted water is used for industry or for supplying drinking quality water to the public (Table 2.2). Public water supply is composed not only of water abstracted for use in households, but also any water abstracted to be used as 'mains' or treated water in factories, businesses, farms etc. The supply of water in England and Wales is a major activity that involves 1584 boreholes, 666 reservoirs and 602 river abstraction points.

Table 2.2 Water abstraction comparison (Source: UK Environment Agency, 2001).

	Total abstraction (cubic km)	%Domestic	%Industrial	%Agricultural
England and Wales	15.26	41	45	14
France	37.73	16	69	15
Germany	46.27	11	70	20
Italy	56.20	14	27	59
Netherlands	7.81	5	61	34
Portugal	7.29	15	37	48
Spain	30.75	12	26	62
United States	467.34	13	45	42
Canada	45.10	18	70	12

Table 2.3 Comparison of water qualities.

Parameter	Type of water					
	Borehole	Moorland	Upland	Lowland	Brackish	Sea
Colour (Hazen)	1.3	136	128	8.25	5	–
Turbidity (NTU)	0.09	1.5	5	22	<0.1	5
Conductivity (μS cm^{-1})	700	–	70	775	2250	51 000
Dissolved organic matter (mg l^{-1})	2	12	15	5.3	–	>1
Algae (cells ml^{-1})	0	0	–	–	–	–
Bacteria (No l^{-1})	1700	0	–		–	–
Total pesticides (μg l^{-1})	0.2	0	0	0.35	–	–
Cryptosporidium (No l^{-1})	0	0	0	0.05	–	–
Nitrate (mg l^{-1} N)	18	–	–	4.8	–	–
Arsenic (mg l^{-1})	–	–	–	3.7	–	–
pH	7.3	6.3	6.5	8.1	7.5	7.9
Hardness (mg l^{-1} CaCO$_3$)		–	2	353	60	350
Sodium (mg l^{-1})	–	–	–	45	257	10 300
Alkalinity (mg l^{-1} CaCO$_3$)		–	20	236	–	170
Chloride (mg l^{-1})	55	–	7	70	502	20 510
Aluminium (mg l^{-1})	0.04	0.09	0.3	0.23	–	–
Iron (mg l^{-1})	0.15	0.42	2.4	0.03	–	–
Manganese (mg l^{-1})	0.0	0.04	0.07	0.0	–	–

2.3   Water sources

The water that is used for drinking water treatment plants and for many industries is categorised, according to its source, as upland surface water, lowland surface water, groundwater, brackish well water and seawater. The characteristics of these waters are compared (Table 2.3) and it is these characteristics that determine the level of treatment required.

Upland surface waters are derived from moorland springs and rivers and because they have had little contact with mineral deposits in the soil they are generally low in minerals (typically less than 150 mg l^{-1} TDS) and are soft. Upland water is typically low in solids but contains significant quantities of dissolved organic matter. The organic content of stream water is seasonally variable and can be very high at the start of the rainy season. Upland waters standing on thin coverings of land over impervious rock are termed moorland waters and as these waters run through peaty earth and lie in peat bogs, they are high in dissolved organic matter (10–20 mg l^{-1}). The organic materials can be arbitrarily divided into fulvic and humic acids (respectively, soluble and insoluble at low pH) and give the water a yellowish or brown colour. The water quality can vary significantly from season to season especially following the first winter rains. Treatment drivers for these water sources are the reduction in organic matter to control colour and disinfection by-product formation and the

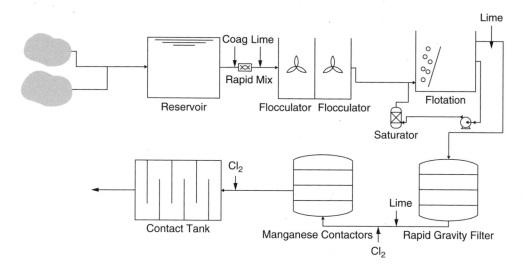

Fig. 2.1   Example flowsheet for an upland catchment reservoir.

removal of metals such as aluminium, iron and manganese. An example treatment flowsheet for an upland reservoir is shown in Figure 2.1.

Lowland rivers are fed from upland lakes and groundwater springs and their characteristics are somewhere between those of ground water and upland water. However, they also receive effluents from municipal sewage treatment works and industrial outfalls, which further affects their chemical and biological characteristics. Water is often abstracted

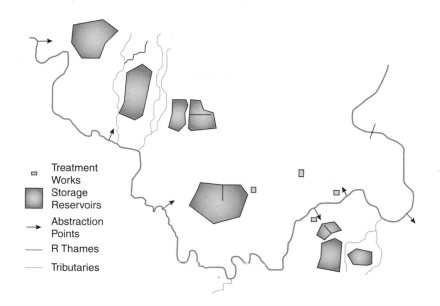

Fig. 2.2   Water supply features in the lower Thames valley (Source: Thames Water).

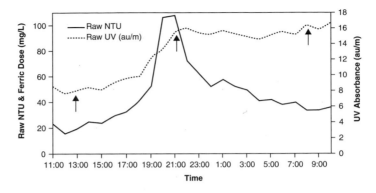

Fig. 2.3 Raw water quality changes during storm events.

from lowland rivers and stored in reservoirs, the water supply for the Lower Thames Valley is shown above (Figure 2.2) where a number of abstraction points from the River Thames feed storage reservoirs and water treatment works.

Lowland waters are also susceptible to seasonal changes in water quality but they are also significantly effected by rapid changes of water following rainstorms. For example the impact of a rainstorm on the water quality entering a water treatment works in the south of England, can be seen in Figure 2.3. The treatment works abstracts water from a lowland river, whose catchment is predominantly agricultural. Raw turbidity increased significantly during rainstorm events and the average dissolved organic carbon (DOC) concentration of the river increased by 40%. The alkalinity of the river water reduced by around 10% as did the conductivity, and pH was also seen to decrease during storm events. All these changes have a significant impact on the treatment processes.

Treatment drivers for these water sources include the removal of pesticides, algal toxins, disinfection by-product precursors as well as nitrate and bromate. In addition, slow moving waters are normally bacteriologically unsafe due to a wide variety of waterborne bacteria and viruses including *Cryptosporidium*. High nutrient loadings lead to widespread algal growth in many of these waters which will affect process selection. An example treatment flowsheet for a lowland river or reservoir is shown in Figure 2.4.

Water which percolates through the soil and accumulates in underground aquifers is called groundwater and, because of its prolonged contact with mineral deposits, may be characterised by high dissolved solids resulting from percolation through deep layers of minerals. Because of their porosity, chalk hillsides, like those found in the south of England, are particularly good sources of groundwater which may emerge as springs. Because these waters have had prolonged exposure to calcium carbonate

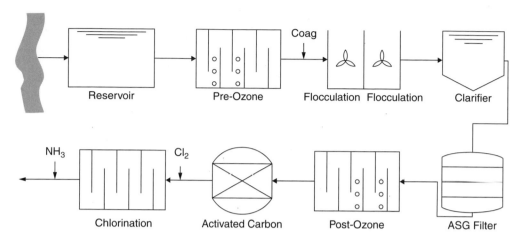

Fig. 2.4 Example flowsheet for a lowland river or reservoir.

and sulphate they are typically hard and alkaline. The total suspended solids (TSS), bacteriological and organic content of groundwaters are usually very low due to percolation through soil and rock, which not only filters out the microorganisms but also adsorbs the organics on which they feed. Seasonal variation is minimal. Treatment drivers for groundwaters include inorganics that can leach from rocks such as iron, manganese and arsenic and often these water sources are impacted by surface water ingress leading to lowland source pollutants such as pesticides, nitrate and *Cryptosporidium* entering groundwaters. An example treatment flowsheet for a groundwater known to have surface water ingress is shown in Figure 2.5.

Brackish waters come from wells in arid areas although some coastal areas have brackish wells as a result of seawater intrusion. Brackish waters are those which have salinities in the range 1000–5000 mg l^{-1}. Seawaters are typically of salinity 20–50 g l^{-1}.

2.4 Water demand

Across Europe around 353 km^3 $year^{-1}$ is abstracted which means that 10% of Europe's total freshwater resources is abstracted on an annual basis. On average, 33% of total water abstraction in European states is used for agriculture, 16% for urban use, 11% for industry (excluding cooling) and 40% for energy production.

Analysis of water demand data identifies at least four components making up its variability; a yearly (seasonal) cycle, a weekly cycle, a degree of randomness and a long-term trend factor (see Figures 2.6 and 2.7). There

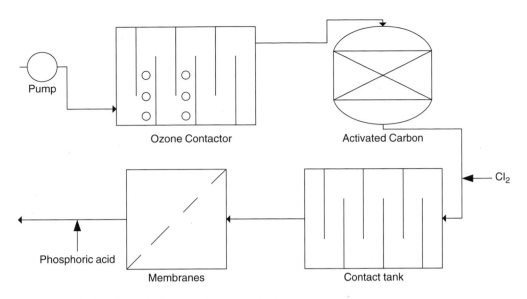

Pump

Ozone Contactor

Activated Carbon

Cl$_2$

Phosphoric acid

Membranes

Contact tank

Fig. 2.5   Example flowsheet for a groundwater.

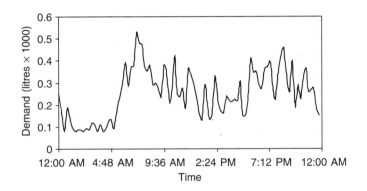

Fig. 2.6   Twenty-four hour demand for a residential neighbourhood in the UK.

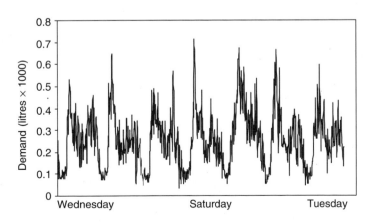

Fig. 2.7   Weekly demand for a residential neighbourhood in the UK.

are a number of factors which influence our demand for water. For example the pattern of demand in the UK has significantly changed over the last 30 years with industrial use decreasing and domestic use nearly doubling. Total abstractions from surface and ground water in England and Wales fell between 1971 and 1993 from 42 700–32 300 Ml day^{-1} but rose to 42 500 Ml day^{-1} in 2002 mainly due to increased demand from the power generation industry. The increase in domestic demand has been driven mainly by the more widespread use of dishwashers, washing machines and garden sprinklers and the increase in one and two person households. Much of the water use in households is for toilet flushing (33%), bathing and showering (20–32%) and for washing machines and dishwashers (15%). The proportion of water used for cooking and drinking (3%) is minimal compared to the other uses. In Europe, the Spanish are the greatest users of water using 265 l/capita/day, followed by Norway (224 l/capita/day), Netherlands (218 l/capita/day) and France (164 l/capita/day). Lithuania, Estonia and Belgium with 85,100 and 115 l/capita/day, respectively, have the lowest household water use. In the UK the current use is around 155 l/capita/day (Downing *et al.*, 2003).

2.4.1   Climate

Climate has a significant overall influence on annual and seasonal patterns of per capita water use. Climatic factors (temperature, precipitation, solar radiation, humidity, wind) operate in combination to influence the amount of water consumed by all sectors, but particularly by agriculture. Their effects differ with location and can fluctuate both seasonally and from year to year. Temperature is the climatic factor that most influences water consumption. Climate change is already affecting rainfall across Europe where some northern European countries have already seen an increase of more than 9% of the annual precipitation per decade between 1946 and 1999 (Downing *et al.*, 2003; IPCC, 2001). Southern and central Europe are predicted to experience a decrease in rainfall. The increasing rates are mainly due to more precipitation during the winter months, while southern Europe will experience more summer droughts.

2.4.2   Income

Cities with a large per capita income generally use more water per capita than cities with lower per capita income. Large residential premises in higher income areas require more irrigation water and water costs tend to constitute a smaller fraction of the household financial budget. Also, higher income families are likely to own more water-using appliances.

2.4.3 Water price

Pricing influences demand by providing an incentive for customers to manage water use more carefully. The customer's sensitivity to pricing may be enhanced by rate structures that increase the cost of water at a constant or increasing rate as use increases. Generally, metering causes a reduction in water use and provides useful planning data for establishing a rate structure. Although there is opposition to metering, the trend toward metering all deliveries in Europe continues.

2.4.4 Water-using appliances and other household uses

More water-using household appliances came into use during the last century. A variety of these water-using or water-operated fixtures were added to homes and early models were not designed to conserve water, contributing to increasing per capita water use trends. Most newer models on the market now require far less water than their predecessors. For example, new horizontal axis washing machines are being introduced that reduce indoor water use further. The introduction of new water-using devices is also diminishing.

2.4.5 Community make-up

Water use trends are affected by changes in the patterns of land use. As cities age and enlarge, the centres are progressively developed for higher population density uses under urban renewal or similar programs. Single-family dwellings are replaced by multiple-unit structures with some land set aside for commercial strips and other industrial development. In small communities, the addition or removal of a single high-water-using entity, such as a food processing plant, can noticeably increase or decrease the per capita water use. The effect can be less pronounced for a major city where one addition or removal of a water-intensive industry may have minimal impact. Many communities have an influx of non-residents (daily, on weekends or seasonally) who are probably counted as permanent residents of another community and are not considered in per capita water use calculations. Relying solely on gross per capita use values in these communities can be misleading.

2.4.6 Effect of tourism on demand

Large influxes of people to a city or region can clearly have a significant impact on the demand for services and many holiday destinations suffer from extreme seasonal variations in demand for water. The Mediterranean region is particularly prone to this sort of phenomenon as populations can expand by a factor of 10 over short periods during peak holiday seasons (particularly on some of the islands). A similar, though less extreme, effect also occurs in the UK where parts of the southwest can experience a doubling of population.

2.4.7 Industrial demand

Experience in industrialised and developing countries alike shows that industries tend to use water more cost-effectively than other sectors of society. While industry is not a large water user compared to other sectors, industries are frequently located in urban areas where water consumption is growing fastest. The price per unit of water is normally set higher for industry than for domestic users, for reasons generally concerned with capital financing infrastructure costs and the higher cost of treating industrial wastewater. Furthermore, industries tend to respond readily to economic and regulatory incentives.

2.4.8 Leakage

Leakage from cracked pipes and losses from service reservoirs is a significant component of overall demand. Leakage can be considered to have two main components (i) bursts having greater than 500 l hr^{-1} flow rate and smaller leaks having from 10 l hr^{-1} up to 500 l hr^{-1} flow rates. The trend in leakage figures for the whole UK water industry is shown in Figure 2.8. In 2002–2003, 3623 Ml day^{-1} of water put into the supply in England and Wales was lost through leakage, 29% lower than in the peak year 1994–1995 but higher than in any of the four previous years. This figure considers the whole industry together but there is considerable regional variation within this depending on the water company. Leakage in the UK is much lower than in many countries such as Italy (30%), Ireland (34%), Bulgaria (50%) and China (84%) (Source: EEA).

As is the case with other networked utilities, peak demand is as significant a concern for suppliers as is total demand. In the UK, the difference between peak and average demand was no more than 21% in 1991. However, projections suggest that this figure could rise to 36% by 2021. If the climate warmed by as little as 1.1°C, this figure could reach 42%.

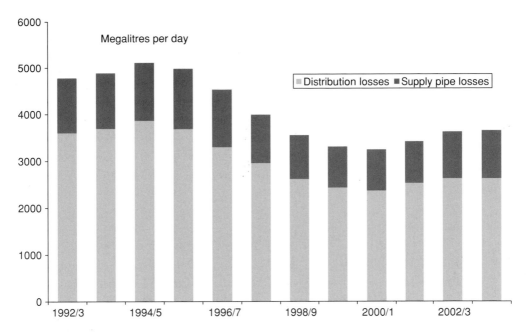

Fig. 2.8   Water leakage in England and Wales: 1992–1993 to 2003–2004, (Source: DEFRA, 2004). Distribution losses include all losses of drinkable water between the treatment works and the highway boundary. Supply pipe losses is leakage from customers' pipes between the highway boundary and the customer's stop tap.

References

DEFRA (2004) e-Digest of Environmental Statistics, http://www.defra.gov.uk/environment/statistics/index.htm, October.

Downing, T.E., Butterfield, R.E., Edmonds, B., Knox, J.W., Moss, S., Piper, B.S. & Weatherhead, E.K. (and the CCDeW project team) (2003) Climate Change and the Demand for Water, Research Report, Stockholm Environment Institute Oxford Office, Oxford.

IPCC (2001) Climate change 2001: The scientific basis. Contribution of Working Group 1 to the Third Assessment Report of Intergovernmental Panel on Climate Change the WMO/UNEP/IPCC.

Coagulation and Flocculation 3

3.1 Introduction

Visible cloudiness was the driving force behind the earliest water treatments, as many source waters contained particles that had an objectionable taste and appearance. To clarify water, the Egyptians reportedly used the chemical alum to cause suspended particles to settle out of water. This is called coagulation and thought to have been used as early as 1500 BC, and is still today one of the most key water treatment stages. The performance of physical separation processes such as sedimentation, flotation and filtration is reliant on the characteristics of the particles they are trying to remove. This is typically in terms of the particle size, shape, and density as well as the particle charge, all of which are controlled to some degree by the coagulation and flocculation processes. The process is a simple but very important stage in most water treatment works and its goal of the agglomeration of fine particles and colloids into larger particles is a well established means of removing turbidity, natural organic matter and other soluble organic and inorganic pollutants.

This agglomeration process is referred to by a number of terminologies that depend on either the application or the mechanism of operation. Terms such as coagulation and flocculation are often used with specific meanings, to refer to different aspects of the overall process. For instance, in water treatment coagulation is often taken to mean the process whereby particles in water are destabilised by dosing certain chemical additives (*coagulants*) and perhaps the rapid formation of small agglomerates. Flocculation is then the process in which destabilised particles and small agglomerates are encouraged to collide with each other (in a stirred or flow-through reactor) to form agglomerates (*flocs*). This is shown schematically in Figure 3.1.

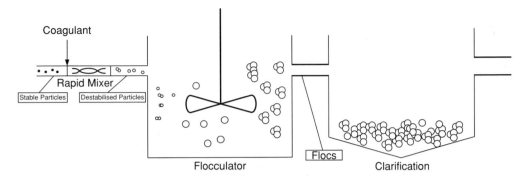

Coagulant

Rapid Mixer

Stable Particles

Destabilised Particles

Flocculator

Flocs

Clarification

Fig. 3.1 Coagulation and flocculation flowsheet.

This chapter will give the reader an understanding of both the process science and application of coagulation and flocculation in water treatment.

3.2 Process science

The sub-micron particles which cause turbidity in water are colloidal, that is they carry a surface charge which makes them repel each other so that they cannot agglomerate into larger particles, which would be easier to remove in any clarification processes (Figure 3.1). This charge is almost invariably negative and is there as a result of a number of mechanisms, the most important being the ionisation of functional groups, carboxylic groups for example, on the surface of the solid and secondly the adsorption of ions or other charged species such as polymers onto the surface.

Direct measurement of the charge on a particle surface is currently not possible and, in the context of this book, is probably not of any significance. A more important characteristic is the electrostatic charge on the surface at the shear plane called the zeta potential (ζ). Most natural colloids in water acquire a zeta potential between -5 and -40 mV due mainly to the presence of charged groups such as carboxylic acids and oxides. As a result of these groups the zeta potential is pH dependent and, in general, as the pH is lowered the charge on a surface increases towards zero. The pH at which the zeta potential is zero is called the iso-electric point (i.e.p.) of the particle and is an important parameter in understanding the nature of the colloidal particles. Acidic groups, like the carboxyl group which generates much of the charge in algae and natural organic matter, dissociate at low pH and hence have a zeta potential which remains fairly stable over neutral and basic conditions. In contrast oxides are amphoteric so that colloidal clays have zeta potentials that vary over both acidic and

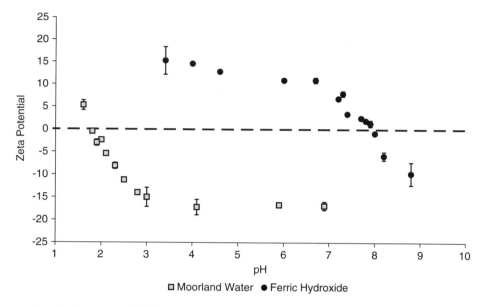

Fig. 3.2   Zeta potential (nV) profiles of a moorland water and a coagulant precipitate.

basic pH ranges. Ferric hydroxide for example is positively charged in acidic waters and negatively charged in alkali waters (Figure 3.2). Typical values of zeta potential for known contaminants are shown in Table 3.1 and a recent survey of UK waters found the majority of raw water particles had zeta potential values ranging from −8 mV to −17 mV.

Table 3.1   Zeta potential of common water contaminants.

Particle	i.e.p.	Zeta potential (mV)	pH
Crushed anthracite	4	−2.8	7
Activated carbon	4	−1.4	7
Kaolin clay	2	−28	7
Humic acid	1.8	−40	4
Fulvic acid	<3	−27	7
Bubbles	2	−26	8
Cryptosporidium oocysts	3.3	−38	7
Giardia cysts	2.2	−17	7
Microcystis aeruginsa	–	−25	9.5

3.3   Coagulation

To remove these stable particles from water it is necessary to neutralise the negative charge. There are a number of mechanisms for destabilising these particles but the most relevant to water treatment rely on the addition of a chemical called a coagulant. These are typically metal salts such as iron

Table 3.2 Common inorganic coagulants used in water treatment.

Chemical	Common name	Formula	Molecular weight	% metal ion	Form	pH	SG
Iron sulphate	Ferric sulphate	$Fe_2(SO_4)_3$	400	14	Liquid		1.58
Iron chloride	Ferric chloride	$FeCl_3$	162	14	Liquid		1.45
Aluminium sulphate	Alum	$Al_2(SO_4)_3 \cdot 14H_2O$	594	17–18% as Al_2O_3	White and greyish powder	2.7	1.34
				8% as Al_2O_3	clear liquid.	3.5	
Polyaluminium chloride	PACL	$Al_2(OH)_5Cl$	n/a	12.4	Clear liquid	4.2	1.34

and aluminium sulphate (Table 3.2). There are a number of coagulation mechanisms including double layer compression, charge neutralisation, sweep flocculation and inter particle bridging; however, the coagulant selected, its dose and water quality determine the mechanism. For instance charge neutralisation is the dominant removal mechanism for waters containing natural organic matter or algae whereas for low turbidity waters sweep flocculation is preferred.

When the coagulants are added to water a series of complex hydrolysis reactions occur. Whilst the coagulants contain Al^{3+} or Fe^{3+}, which are highly charged counter-ions, for most particles these species only exist under acidic conditions (low pH). At the range of normal water pHs these ions react to give various hydrolysis products such as $Al(OH)^{2+}$ and $Al_{13}O_4(OH)_{12}^{7+}$ ('Al_{13}' *polymer*), which can strongly adsorb on negative particles and so reduce the particle charge in this way. In most cases, precipitation of insoluble hydroxides, $Al(OH)_3$ and $Fe(OH)_3$ also occurs.

$$Al_2(SO_4)_3 \cdot 14H_2O \Leftrightarrow Al(H_2O)_6^{3+}$$
$$\Leftrightarrow Al(OH)(H_2O)_5^{2+}$$
$$\Leftrightarrow Al_{13}O_4(OH)_{24}^{7+}$$
$$\Leftrightarrow Al(OH)_3(s) \Leftrightarrow Al(OH)_4^{-}$$

Precipitation of hydroxides also plays a key role in practical flocculation. Large numbers of colloidal hydroxide particles are produced which flocculate to form gelatinous flocs, which enmesh most of the original particles in the water. This is the sweep flocculation mechanism and is very important when the particle concentration is quite low (i.e. for low turbidity waters). Under these conditions, particle collision rates are low and hence flocculation, even for fully destabilised particles, would be slow.

The formation of many new particles (hydroxide precipitates) gives a considerable enhancement of the flocculation rate. The charge neutralisation of stable particles is rapid and can be achieved at quite low coagulant dosages, but the dosage should be proportional to the contaminant concentration. Sweep flocculation should occur at higher dosages and is much slower, but the required dosage should not be greatly dependent on particle concentration.

The different mechanisms outlined above have led to the definition of four zones of coagulant dosage, with the following consequences for negatively charged particles (Duan & Gregory, 2003):

Zone 1: Very low coagulant dosage; particles still negative and hence stable.

Zone 2: Dosage sufficient to give charge neutralisation and hence coagulation.

Zone 3: Higher dosage giving charge neutralisation and restabilisation.

Zone 4: Still higher dosage giving hydroxide precipitate and sweep flocculation.

3.4   Flocculation

Once the coagulant has been added the particles become rapidly destabilised. This leads to the formation of unstable micro-flocs ranging from 1 to around 10 μm. Floc growth occurs due to collision with other flocs until a steady state floc size distribution is reached. These collisions can also lead to the break-up of flocs. Figure 3.3 shows flocs formed during the coagulation of natural organic matter and of kaolin clay.

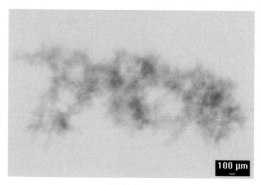

NOM Floc

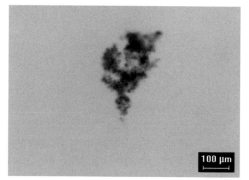

Kaolin floc

Fig. 3.3 Microscope images of typical water treatment flocs.

The agglomeration of colloidal particles into flocs occurs as a result of a number of basic steps that can be summarised into two mechanistic events:

- The particles must collide with one another by either forced or free motion (*collision/transport*).
- The particles must then adhere to one another and remain as a single cluster of particles (*attachment*).

In many ways these two actions, which are usually referred to as transport and attachment, respectively, can be regarded as independent actions and must be treated separately. This is a valid assumption as the range of action of the two steps is generally at least an order of magnitude different. In particular the range of colloidal interactions is very short, typically considerably less than the size of a particle. Whereas, the transport step has to bring the two particles close together from relatively large separation distances over which the colloidal interaction forces do not exert any influence.

The most common method for analysing the transportation stage of the agglomeration process was first introduced by Smoluchowski in 1917. In his approach the system is described in terms of a second-order rate process whereby the rate of collision is proportional to the product of concentrations of the two colloidal species. For simplicity it is assumed the particles are identical and hence the square of the concentration ($N^2 = n_i n_j$). Each time two particles collide to form an agglomerate there is a net loss of *one* particle and the rate of flocculation can be written as the rate of reduction of particle (number) concentration:

$$-\left(\frac{dN}{dt}\right) = \alpha k_F N^2$$

where N is the particle number concentration, α is the *collision efficiency factor* (fraction of collisions leading to agglomeration) and k_F is a flocculation rate coefficient. The exact nature of the rate constant depends on the controlling mechanism. In practical situations there are three significant mechanisms:

- Perikinetic flocculation (diffusion)
- Orthokinetic flocculation (fluid motion)
- Differential sedimentation.

3.4.1 Perikinetic flocculation

All particles dispersed in fluid undergo random displacements, known as Brownian motion due to thermal energy of the system. The rate of perikinetic collision is calculated by determining the rate of diffusion of

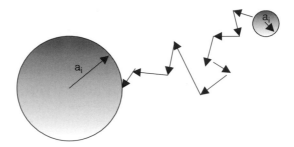

Fig. 3.4 Perikinetic transport.

spherical particles of size a_i to a fixed sphere of size a_j (Figure 3.4, Table 3.3). The impact of particle size is shown in the term $(a_i + a_j)^2/a_i a_j$ which is the amalgamation of the particle size effects with respect to the collision radius $(a_i + a_j)$ and the diffusion coefficient $(a_i + a_j)/a_i a_j$. When the particle sizes are equal the size term equates to a value of 4 and remains fairly constant when the two sizes differ by less than a factor of 2. The resultant value of the perikinetic rate constant for water dispersions at 25°C is 1.23×10^{-17} m^{3}s^{-1}. As the difference in the size of the two particles exceeds a factor of 2 the rate constant starts to increase. For instance when the particles are an order of magnitude different in size the rate constant increases by roughly a factor of 3 to 3.8×10^{-17} m^{3}s^{-1}. Further, as the size of the agglomerate increases its shape is likely to become irregular due to dendritic growth. Under these conditions the increase in the collision radius will more than compensate for the decrease in diffusion coefficient and the rate constant will increase further.

3.4.2   Orthokinetic flocculation

Particle transport induced by forced fluid motion is known as orthokinetic collision and generally results in a significant increase in the rate of inter-particle collisions. The mechanism of capture is due to an imposed shear brought about by fluid flow (natural systems) or by mechanical stirring

Table 3.3 Comparison of rate constants for the different mechanisms of collision.

Mechanism	Rate constant (m$^3 \cdot$s^{-1})	Key feature (ignoring particle size)
Orthokinetic	$\left(\dfrac{4}{3}\right) G(a_i + a_j)^3$	Shear rate (G)
Perikinetic	$\left(\dfrac{2kT}{3\mu}\right) \dfrac{(a_i + a_j)^2}{a_i a_j}$	Temperature (T)
Differential setting	$\left(\dfrac{2\pi g}{9\mu}\right)(\rho_s - \rho)(a_i + a_j)^3(a_i - a_j)$	Density difference ($\rho_s - \rho$)

(engineered systems). Here the rate of collision depends on the size of the particles and the fluid velocity gradient or shear rate (G). In fact, the rate of agglomeration is proportional to the cube of the collision radius such that the larger the particle the more likely it is to collide with other particles. To compare to perikinetic agglomeration the two rate constants are equal when the particle sizes are both 1 μm and the shear rate is 10 s^{-1}. When either the shear rate or the particle sizes increase, the orthokinetic rate rapidly exceeds the perikinetic rate and it is the orthokinetic rate that we can most control. Overall, if the particle size is sub-micron perikinetic collision will be appreciable such as with primary clay particles but becomes relatively unimportant as the particle size increases to beyond 1 μm. Above 1 μm orthokinetic collision mechanisms begin to become significant with the rate of collision increasing very quickly as the particle size increases. Above 3 μm differential settling begins to become important and starts to control the process at particle sizes above 30 μm.

The concept of the average hydraulic gradient, G (s^{-1}) was developed by Camp and Stein in 1943 to characterize mixing in flocculation basins, and the G value is now used worldwide to characterize mixing in a wide range of environmental engineering applications. Camp and Stein termed this the absolute velocity gradient and related this to work done per unit volume per unit time via

$$G_{abs} = \sqrt{\frac{P/V}{\mu}} = \sqrt{\frac{\varepsilon}{\nu}}$$

where

> P is power dissipated
> V is tank volume
> μ is dynamic viscosity of the water
> ε is the power input or energy dissipation rate per unit mass
> ν is the kinematic viscosity.

The absolute velocity gradient can be calculated at any point within a mixing vessel, provided that the power dissipated is known at that point. However, the dissipated power will vary from point to point within the mixing vessel, and consequently the velocity gradient is a function of both time and position. Workers have traditionally replaced the absolute velocity gradient with an approximation of the exact value; that is its average value throughout the vessel, G_{ave}:

$$G_{ave} = \sqrt{\frac{P_{ave}}{V\mu}}$$

where the average power consumption, P_{ave}, is readily obtained via

$$P_{ave} = N_p \, \rho \, N^3 D^5$$

It is this G_{ave} value which has been used as a cornerstone in the design of mixing processes, including flocculators. A generalized flow diagram for a ferric chloride chemical addition system is shown in Figure 3.5. The coagulant is stored in a tank made of either fibreglass-reinforced polyester or rubber-lined steel. A flow meter installed along the main water line is used to pace the addition of the coagulant to the water flowrate. After the coagulant is added the water is mixed with an inline mixer and then passes onto the flocculation stage.

The required Gt is dependent on the water source, typical values for a low turbidity water are between 20 and 70 s^{-1} and for a high turbidity water a Gt of between 50 and 150 s^{-1} is suitable. Examples of hydraulic and mechanical flocculators are shown in Figures 3.6 to 3.8.

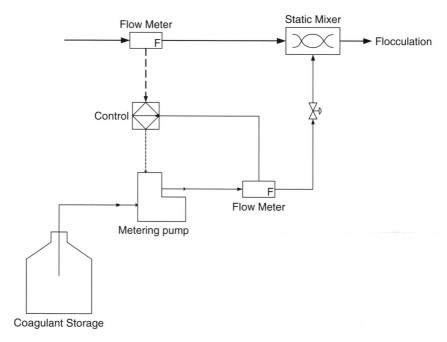

Fig. 3.5 Coagulant addition flow diagram.

3.5 Applications

Coagulation and flocculation are used as the key process for removing suspended particles and dissolved inorganic and organic contaminants. The

Fig. 3.6 Example of a static mixer used for blending water sources and coagulation at an upland reservoir site.

nature of the contaminant load varies from source to source. For example, source water originating from rivers can have a high proportion of suspended clay colloids, whereas upland, peaty areas are generally dominated by NOM. In all source waters algae are ubiquitous, although abundance differs depending on the extent of eutrophication. Seasonal algal growth can interfere extensively with a process that has been optimised for either clay or NOM systems. This typically results in algal and coagulant carry

Fig. 3.7 Examples of hydraulic mixers used for (i) rapid mixing and (ii) flocculation at a lowland reservoir treatment works.

Fig. 3.8 Example of a paddle flocculator used at a lowland river site.

over, increased coagulant demand and filter clogging. Table 3.4 compares the characteristics of the three main types of water treatment contaminants.

The coagulation process is generally optimised for a particular system in terms of coagulant dose and pH, achieved through a series of bench scale jar tests. Figure 3.9 shows the results of jar test and zeta potential measurements for kaolin suspensions with alum. Below 1 mg l^{-1} as alum there is no reduction in turbidity and the particles are negatively charged (-20 mV) and stable (Zone 1). There is a narrow range of low turbidity when the alum dose is between 1 and 3 mg l^{-1}, which is close to the dosage where the zeta potential is reduced to zero (Zone 2). As the dose of Al increases, the particles become positively charged ($+10$ mV) and re-established, turbidity returns to that of the original kaolin solution (Zone 3). Beyond the dose of 25 mg l^{-1} the turbidity falls again as a result of sweep flocculation (Zone 4). Here there is a reduction in turbidity even though the particles are still positive and even though the alum dose is increased the particle

Table 3.4 Characteristics of NOM, algal and kaolin systems (adapted from Henderson *et al.*, 2005).

	NOM	Algae	Kaolin
Concentration	8.8–14 mg l^{-1}	5×10^5 cells ml^{-1}	50 mg l^{-1}
Turbidity (NTU)	5.9–7	3.2	50
Particle size (μm)	0.15 ± 0.02	4.5	0.2
Density (g cm^{-3})	1.00	1.07	2.67
Specific surface area (m^2 g^{-1})	40	1.09	9.09
Charge density (meq g^{-1})	10–15	Variable	0.1–1
Zeta potential (mV)	-18 at > pH 3	-15 at > pH 3	-50 at > pH 6
i.e.p.	1.5	1.5	2
Optimum coagulation pH	4.5	5	7

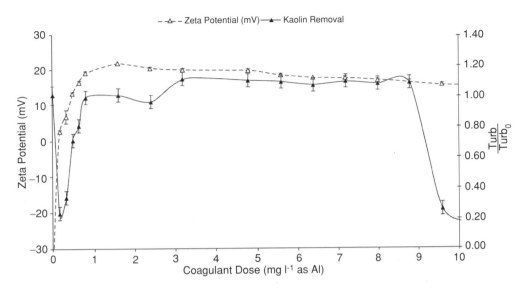

Fig. 3.9 Coagulation of kaolin clay.

charge does not change. The residual turbidity in Zone 4 is significantly lower than in Zone 2, indicating for this example here a much greater degree of clarification by sweep flocculation. The equivalent plots are shown in Figures 3.10 and 3.11 for natural organic matter (NOM) and algae, respectively. For both systems the four zones can be clearly observed.

The dose of coagulant required to remove inorganic and organic contaminants is very different. Organic particles such as NOM and algae

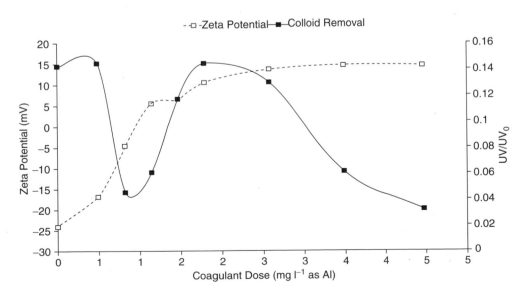

Fig. 3.10 Coagulation of natural organic matter.

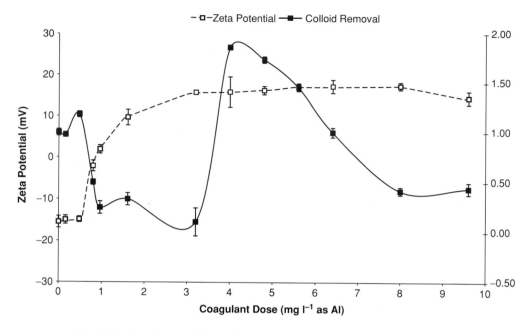

Fig. 3.11   Coagulation of algae (*Chlorella sp.*).

required a colloid to coagulant weight ratio of between 1 mg mg^{-1} and 10 mg mg^{-1} to reach the i.e.p., whilst the ratio required for kaolin was 100 times greater (Figure 3.12). The difference between kaolin and NOM can be explained in terms of the specific surface area (SSA) and charge density, as both parameters are far greater for NOM (Table 3.4). However, with respect to algae, the relatively low SSA (1.09 m^2 g^{-1}) does not fit the observed data despite the high charge density but it is known that algae excrete extracellular organic matter (EOM) which is made up of neutral and acidic components such as polysaccharides and uronic acids. These species are known to interfere significantly with the coagulation process.

Despite the differences between each of the three systems and their coagulation mechanisms, zeta potential has been shown to be a useful parameter in determining the appropriate coagulant dose and pH. It allows us to link the fundamental surface characteristics of the particles to their removal efficiency. Henderson *et al.* (2005) showed how the removal of kaolin, NOM and algae was greater as the magnitude of the zeta potential was reduced, resulting in optimum operational zeta potential ranges (Figure 3.13). The definition of a range, as opposed to a single value, suggests that in general, colloidal destabilisation occurs before complete charge neutralisation of surface charge which is why good optimal residuals are still possible although the zeta potential may not necessarily be zero.

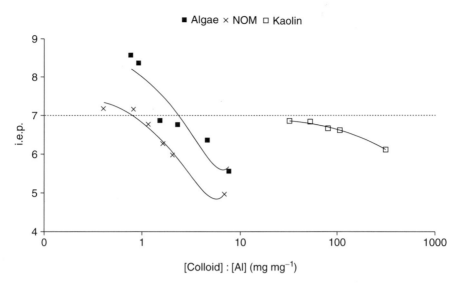

Fig. 3.12 The colloid to coagulant weight ratio versus the i.e.p. for kaolin, algae and NOM systems (adapted from Henderson *et al.*, 2005).

The extent of the range is likely to be determined by additional non-DVLO forces present in the system, such as hydrophobic or steric effects, and is therefore likely to be dependent on the character of the organics present.

Henderson *et al.* (2005) found that the operational ranges were different depending on the contaminant being removed. NOM was successfully

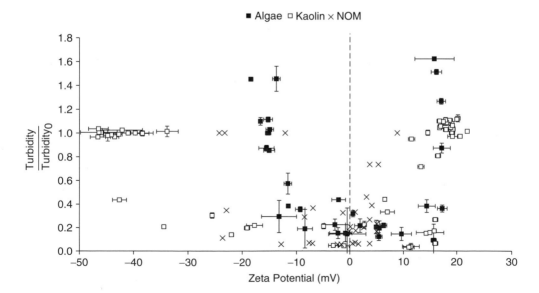

Fig. 3.13 The correlation between zeta potential and removal efficiency of impurity particulates kaolin, algae and NOM systems (adapted from Henderson *et al.*, 2005).

removed between −10 mV and +5 mV, whereas the zeta potential band for kaolin was much wider at −20 mV to +5 mV. The algae had a symmetrical optimal removal band of −12 mV to +12 mV, which was more like the removal band for NOM than for kaolin. This implies that the organic particles are much more reliant on charge neutralisation for removal than inorganic particles. Optimum conditions for organic particles required a low pH (∼pH 5–6) where the dominant removal mechanism is charge neutralisation. However, for inorganic particles coagulation was conducted at pH 7. At this pH, not only would charge neutralisation occur to a degree, considered to be a result of physical adsorption of the cationic amorphous hydroxide onto the surface of the inorganic particle (Duan & Gregory, 2003), but sweep flocculation would also occur increasing the density of the flocs.

3.6   Test methods

3.6.1   Jar tests

In practical processes it is very important to determine the optimum dosage of coagulants and to establish optimum chemical conditions (e.g. pH). For this purpose, the well-known jar test procedure is very widely used in water treatment laboratories (Figure 3.14). Samples of water are dosed with varying amounts of coagulant under standard rapid mix conditions, in order to distribute the additive evenly. The samples are then given a standard period of slow stirring at a fixed speed, during which flocs may form. A certain period of sedimentation is then allowed, after which samples of water are withdrawn for turbidity measurement. The *residual turbidity* of the settled samples gives a good indication of the degree of clarification obtained and can be used to locate the optimum flocculation conditions. The tests are normally conducted in a purpose-built apparatus with multiple stirrers (Figure 3.14), so that a number of samples (usually six) can be tested simultaneously. A typical procedure may involve 1 min of rapid mix at 300 rpm, 15 min of slow stirring at 30 rpm and 15 min sedimentation. With a varying raw water quality, the test may need to be carried out quite often in order to ensure that optimum flocculation conditions are maintained.

3.6.2   Zeta potential

The fundamental principle of the measurement of zeta potential is the measurement of the velocity of particles undergoing electrophoresis, which is

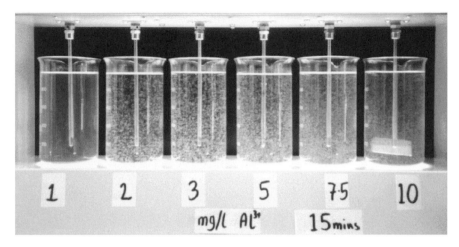

Fig. 3.14 Jar testing experiment.

the movement of a charged particle, under the influence of an applied electric field, relative to the liquid in which it is suspended. The sample is introduced into a horizontal electrophoresis cell with electrodes at either end. When a known electrical field is applied across the electrodes, particles with a negative zeta potential will migrate towards the positive electrode and vice versa. The velocity is a function of the particle's charge, the viscosity of the medium and the applied field. The particle size range for zeta potential measurement is between 5 nm and 30 μm and these particles, which have a very low inertia, will reach terminal velocity in microseconds.

As the charged particles move through the measurement cell and cross focused laser beams the particles scatter light with a frequency that is proportional to velocity expressed as electrophoretic mobility. The zeta potential is derived from the electrophoretic mobility using the Henry equation:

$$U_E = \frac{2\varepsilon\zeta f(k, a)}{3\eta}$$

where

U_E: electrophoretic mobility
ε: dielectric constant
ζ: zeta potential
$f(k, a)$: Henry's Function
η: viscosity.

References

Duan, J. & Gregory, J. (2003). Coagulation by hydrolysing metal salts. *Adv. Colloid Interface Sci.*, 100–102, 475–502.

Henderson, R., Sharp, S., Jarvis, P., Parsons, S.A. & Jefferson, B. (2005). Identifying the linkage between particle characteristics and understanding coagulation performance. *Water Sci. Technol.*, in press.

Clarification Processes 4

4.1 Introduction

Many impurities in water are in the form of suspended particles which have a density greater than that of water. They remain in suspension due to the turbulent motion of water as it flows. However, if the conditions are made more quiescent in nature, these particles will settle out under the influence of gravity. This is the principle behind clarification processes. Typical examples of particles removed in this way include silt, mud, algae and materials that are converted into suspended flocs through the coagulation process such as natural organic matter.

The application of clarification dates back to the early historical records when humans stored water in jars and decanted the top layers. The first large scale settling systems were constructed by the Romans in the form of the castellae and piscinae of the aqueduct systems. This was taken to the next stage during the start of the industrial age where settling reservoirs were used to store large volumes of water. The use of settling basins led to the development of constructed rectangular tanks and then radial flow devices. The first reported developments of high rate sedimentation in the form of lamella plates dates back to Sweden in the 1960s from a need to house the treatment units against the winter extremes. The most recent developments have been the use of ballast to enhance floc densities with its origins in Australia and France.

This chapter will cover the process science and design of sedimentation processes as used in potable water treatment and will follow with a description of the main types of processes used including typical performance data.

4.2   Process science

Particles may settle in one of four distinctively different ways depending on concentration and the relative tendency of the particles to agglomerate while they settle. At low solid concentrations, typically less than 500–1000 mg l^{-1}, settlement occurs without interference from neighbouring particles. As the concentration increases the influence of surrounding particles increases the settling rate. As the particle concentration increases further the process changes from clarification to hindered settling and thickening, which will not be discussed here.

As a discrete particle settles it will accelerate, under the force of gravity, until the drag force on the particle balances its weight force. At this point the particle descends at a constant velocity called the terminal settling velocity. The exact expression for the terminal settling velocity depends on the flow regime around the particle as it settles. However, in most cases in potable water treatment the particles fall in a laminar flow field and the expression becomes the well-known Stokes' law:

$$v_t = \frac{g\left(\rho_p - \rho\right)d^2}{18\mu}$$

where v_t is the terminal settling velocity (m s^{-1}), ρ_p is the particle density (kg m^{-3}), ρ is the density of water (kg m^{-3}), d is the particle diameter (m) and μ is the viscosity of water (kg m^{-1} s^{-1}). The above expression shows the importance of the density and viscosity of water in the settling of particles (Table 4.1). The key issues relate to temperature which has a dramatic impact on the viscosity of water such that particles settle faster in warmer water such that it is possible for the rate to double between summer and winter.

The settling rate of flocs is complicated as they are not solid spheres such that the settling rate changes as a function of size, structure and density. Standard settling rates are usually reported in relation to sand due to its consistent properties and availability. Typical settling rates of sand (specific gravity of 2.65) are 100 mm s^{-1}, 8 mm s^{-1} and 0.154 mm s^{-1} for

Table 4.1 Properties of water and their impact on settling.

T (°C)	ρ (kg m^{-3})	μ (kg m^{-1} s^{-1})	Relative settling rate to 10°C (%)
10	999.73	1.3097×10^{-3}	–
20	998.23	1.0087×10^{-3}	130
30	995.68	0.8007×10^{-3}	164
40	992.25	0.6560×10^{-3}	200
50	988.07	0.5494×10^{-3}	238

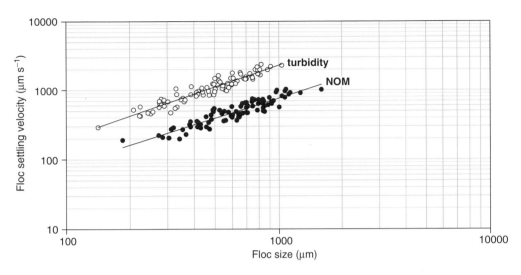

Fig. 4.1 Examples of settling rates for NOM and turbidity derived flocs (adapted from Jarvis *et al.*, 2004).

particle sizes of 1000 µm, 100 µm and 10 µm, respectively. Flocs normally treated in sedimentation processes include NOM and turbidity based agglomerates. Typical settling rates over the sizes they are likely to form are between 0.2 and 1 mm s^{-1} for NOM and 0.5 and 2.5 mm s^{-1} for turbidity flocs (Figure 4.1).

4.2.1   Settlement in tanks

The easiest separation process to be developed for the continuous clarification of particles from water is a horizontal settling tank (Figure 4.2). Consider a tank L m long, D m deep and W m wide which is operated at a flow rate of Q m^3 hr^{-1}. A particle at the inlet to the tank must fall to

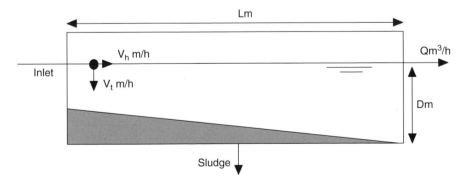

Fig. 4.2 Schematic of a horizontal settling tank.

the bottom of the tank in the time it takes the flow to exit the tank to be removed. The time taken to fall the depth of the tank is

$$t = \frac{D}{v_t}$$

And the time for the water to travel the length of the tank is

$$t = \frac{L}{v_h} = \frac{L}{Q/(D \times W)}$$

Combining the above expressions yields

$$v_t = \frac{Q}{(L \times W)}$$

Where $L \times W$ is the cross-sectional area of the tank such that $Q/(L \times W)$ is the surface loading rate and represents the key design parameter for sedimentation. If the actual settling velocity is lower than the hydraulic loading rate the particle will be entrained in the flow and not be retained. An important consequence of the above is that in theory the depth of the settlement tank is not important in deciding the performance of a sedimentation process. In reality, issues to do with flow stability and scouring mean that depth does play an important role.

4.3   Technology options

4.3.1   *Horizontal flow clarifiers*

Horizontal clarifiers are based on either rectangular or circular (Figure 4.3) configurations and were at one time the standard clarification process around the world. However, they are rarely used now in the UK and Europe for new works. They are still used in North America due to their ease of covering and construction. In rectangular tanks the water to be treated flows in at one end and exits from the other. Typical surface overflow rates are in the range 1–2 m hr^{-1} for coagulated water (typically closer to 1 m hr^{-1}) with retention time of 2–3 hr. The key design issues relate to good flow distribution at the inlet with minimal disturbance to prevent floc break-up and the length to width ratio. Ideally plug flow conditions are required which need a length to width ratio of 20. In practice ratios above 5 tend to be sufficient. Collected sludge is scrapped along the floor of the tank towards a sludge collection chamber near the inlet end of the tank and pumped away for further processing. Circular clarifiers operate from a central feed well with the flow moving outwards to the peripheral weirs such that the velocity progressively decreases. The tanks generally contain a sloped floor which is swept by a scraper towards a central collection

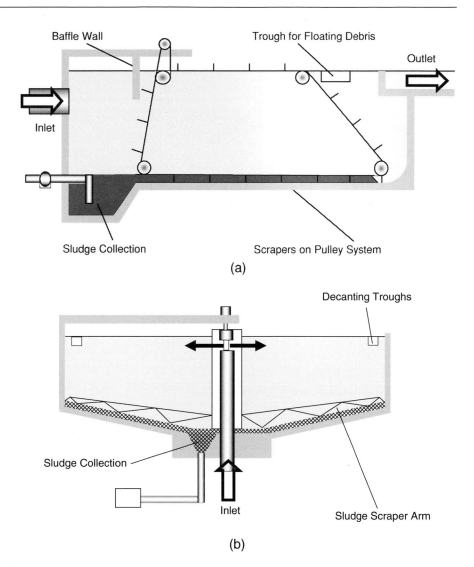

Fig. 4.3 Schematic of horizontal flow clarifiers in (a) horizontal or (b) radial configuration.

chamber. The decreasing velocity profile permits slightly higher overflow rate than rectangular tanks and slightly shorter residence time.

4.3.2 Lamella plates

The principal problem with horizontal clarifiers is the space they take up in generating the required area. One method of improving this is to insert a series of plates or tubes at an inclined angle to effectively increase the area for settlement, known as lamella plates (Figure 4.4). There are a number of proprietary designs (Figure 4.5) based on either circular, hexagonal

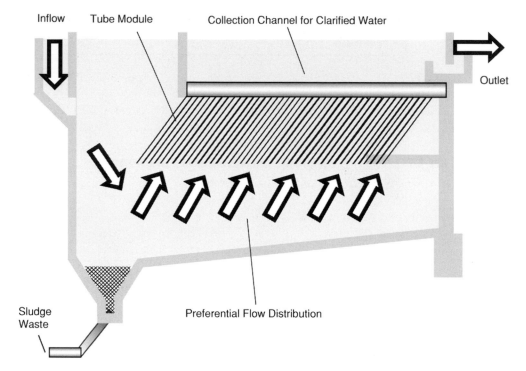

Fig. 4.4   Schematic of a lamella plate clarifier.

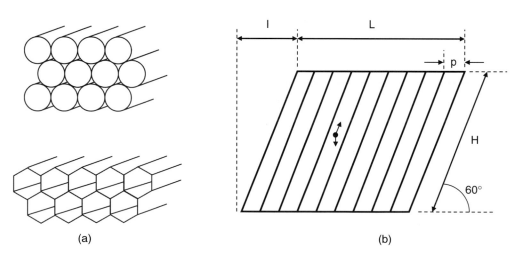

(a)   (b)

Fig. 4.5   (a) Various lamella formats (adapted from Gregory *et al.*, 1999); (b) Schematic of a lamella plate pack.

or rectangular tubes but most have a spacing of 50 mm and a length of between 1 and 2 m. The plates are pitched at an angle of between 50° and 70° so that they are self cleaning. Lower pitches are occasionally used down to as low as 7° but these require periodic back flushing to keep the channel from blocking. In most clarification processes the water flows up through the plates. Particles are separated as they fall onto an individual plate or tube and slide down until they are collected as sludge.

Considering a lamella plate pack with N plates each W wide and H long set at an angle of 60° on a pitch of p (Figure 4.6) the total surface area available for settling can be calculated as

$$A_{effective} = W \times (Np + \cos 60°)$$

Inclusion of the plates within horizontal flow or floc blanket clarifiers can permit an increase of loading rate of the order of 2–3 without deterioration in the effluent concentration. Typical loading rates for such systems are between 5 and 10 m hr^{-1}.

4.3.3   *Vertical flow clarifiers – sludge blanket clarifiers*

The floc or sludge blanket clarifier is a simple design that comprises a circular or square hopper bottom tank (Figure 4.6). Coagulated water is fed to the apex of the hopper and the water flows in an expanding upflow (decreasing velocity) pattern that allows flocculation to occur. At a point designed to be around the top of the hopper section, the upflow velocity of the water equals the settling rate of the flocs such that they remain in suspension with clean water above. The suspended flocs accumulate and form a blanket which acts as a filter bed capturing small flocs and particles such that the clarification process is enhanced. As flocculated particles continue to accumulate in the blanket its height increases towards the top of the tank. The level of the blanket is controlled by removing solids to keep a clear water zone between the blanket and the outlet weirs.

The ability of the blanket to capture small flocs and particles is dependent on the blanket concentration, the flux of solids into the blanket and the upflow velocity (Figure 4.7). The effluent turbidity deteriorates rapidly once a certain threshold upflow velocity has been exceeded (point A) as the floc blanket concentration (point B) is decreased below the point of maximum flux into the blanket (point C). Figure 4.7 also demonstrates that there is no improvement in turbidity in increasing the blanket concentration beyond that obtained at the maximum particle flux as the blanket is working at its maximum capture efficiency.

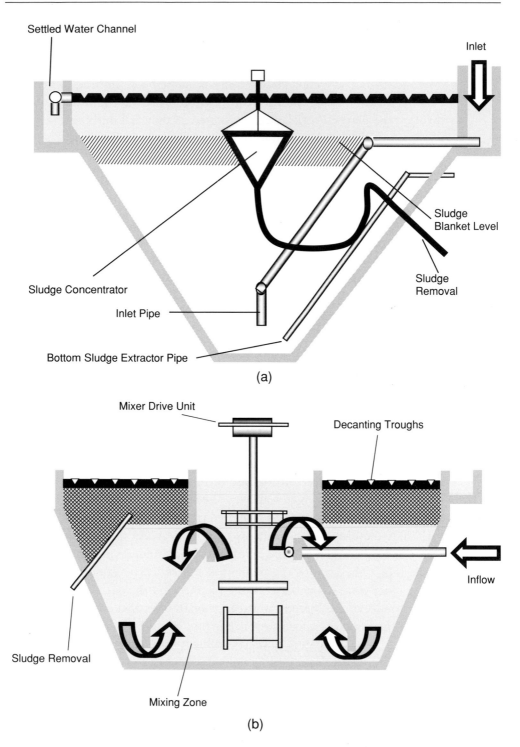

Settled Water Channel

Inlet

Sludge
Blanket Level

Sludge Concentrator

Sludge
Removal

Inlet Pipe

Bottom Sludge Extractor Pipe

(a)

Mixer Drive Unit

Decanting Troughs

Inflow

Sludge Removal

Mixing Zone

(b)

Fig. 4.6   Schematic of floc blanket clarifiers – (a) hopper bottom and (b) flat bottom.

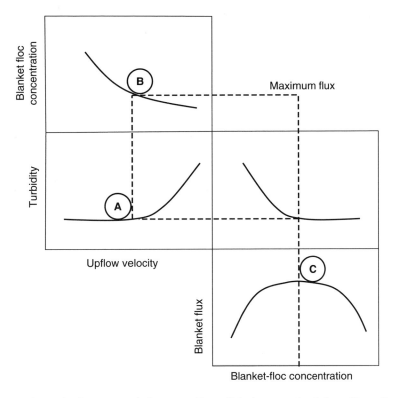

Fig. 4.7 Relationship between settled water quality and blanket operation (adapted from Gregory *et al.*, 1999).

A very stable floc blanket can be operated with only a small clear water region above it (0.1 m) although slightly deeper levels are often used to compensate for process instabilities. Upflow velocities are set at about 50% of the terminal settling velocities of the free particles which equates to around 1–3 m hr^{-1}. Higher flow rates are possible with water softening plants which can be operated as high as 5 m hr^{-1}. The drawback of the design is that a reasonably deep tank is required to generate the hopper design, as the hopper angle is around 60°, to avoid sludge sticking to the walls. This sets a practical limit to the size of an individual tank to about 5 m.

To enable larger tanks a number of modified designs have been developed including multi-hopper and flat bottom tanks. The latter are more common in new plants which use either multiple inverted candelabra feed pipes or laterals across the floor. One version of the process is the Pulsator clarifier (Superpulsator®), made by Infilco Degremont, which derives its name from the fact that the inflow enters the tank in pulses (Figure 4.8). Incoming water enters through a chamber where the water level is raised

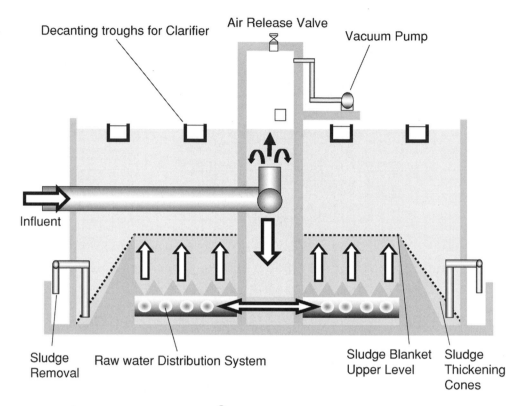

Fig. 4.8 Schematic of a Pulsator® clarifier.

above the water level in the tank under the influence of a partial vacuum. When the vacuum is released, by opening a vent, the flow surges out through the distribution pipework. Consequently, there are cycles of low flow followed by a quick pulse. A typical cycle lasts around 30–60 s during which time the sludge blanket will contract (surge) and expand (low flow) which generates a reportedly more uniform blanket. Widely spaced inclined plates are also included to suppress currents and enhance settling which results in an increase in the loading rate to between 1.2 and 12 m^3 m^{-2} hr^{-1}.

4.3.4 High rate systems

Sand ballasted flocculation (Actiflo®)

The Actiflo® process uses fine sand in the size range of around 80–160 μm as a ballast to enhance the settling rate of the flocs and increase the growth kinetics during flocculation (Figure 4.9). Settling rates are observed to be an order of magnitude higher than in traditional systems with a

Fig. 4.9 Photograph of a ballasted floc (with kind permission of OTV – Veolia Water Systems).

flocculation process which is 2–8 times quicker. The combined benefits mean that the process is a small footprint clarification process with about a 60–80% reduction in area compared to a traditional settler. The process was first introduced in the early 1990s for drinking water production from river water sources in France. By the year 2004, total installed capacity for drinking water treatment was 10 340 160 m^3 d^{-1}. The majority of the plants are in the USA and France, they range in size from 30–25 000 m^3 hr^{-1} and are still based on river water sources. The technology is also used widely for wastewater and storm water/CSO applications where there is a total installed capacity of 10 681 800 m^3 d^{-1} by the end of 2004.

The process includes three separate stages (Figure 4.10). Raw water and coagulant enter the first stage, *injection*, where the micro sand and a polymer are injected into a mixing zone to combine the ballast to the pollutants in the water. Typical retention times in the zone are of the order of 2 min. The flow then passes to the second stage, the *maturation* zone, where the flocs grow in a reduced shear environment of typically 160–200 s^{-1} for a further 6 min. The flocs then pass to the final settlement zone which incorporates lamella plates. Residence time in this zone is of the order of 2–3 min resulting in a total residence time in the process of about 10 min. The settled mixture of sand and sludge is pumped to a hydrocyclone where the sand is separated from the sludge and recycled. The sludge flow typically accounts for 6% of the total incoming flow (Plum *et al.*, 1998).

The effectiveness of the process can be seen in a simple jar test (Figure 4.11) which shows the speed of the process. In batch situations settling

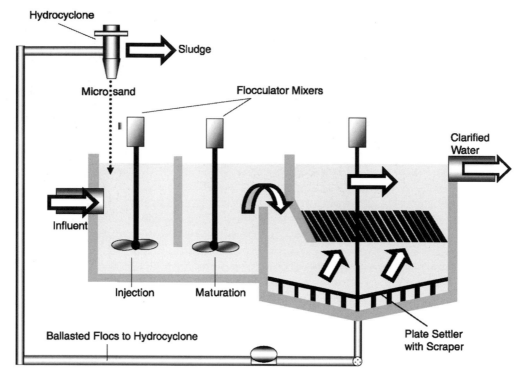

Fig. 4.10 Schematic of the Actiflo® process.

times of 10–30 s are adequate for clarification although the bulk of the settlement has been reported to occur within the first 3–5 s. The main design parameters are the overflow rate and the dosage rates of sand and polymer. Overflow rates of up to 200 m hr^{-1} have been suggested but the majority of plants operate in the range 40–80 m hr^{-1}. Sand dosage rates maintain a concentration of 2–4 g l^{-1} and polymer doses are matched to the system characteristics but are typically less than 0.5 mg l^{-1}. The technology has been shown to match and even improve on the removal seen in traditional systems where up to 99% suspended solids removal is possible. One specific study showed that the system produced a robust effluent concentration of around 6 mg l^{-1} irrespective of the load onto the plant (Plum *et al.*, 1998).

Densadeg®

An alternative process uses recycled sludge as the ballasting aid and is used principally for river water intakes (Figure 4.12). The process contains three zones made up of a reactor, thickening and clarification zone. In the reactor zone, pre-formed micro-flocs combine with recycled sludge from the

Fig. 4.11 Stages of operation: Coagulation – 30s; Flocculation – 1 minute; Setting – 10s. (with kind permission of OTV – Veolia Water Systems).

thickening zone in combination with an anionic polymer. The combined flow is drawn through a draft tube at a rate of up to 10 times the incoming feed flow producing compact dense flocs. The flow then passes to the pre-settling/thickening zone where the bulk of the separation occurs. A slow moving rake at the bottom of the tank aids further thickening by releasing entrained water which is periodically pumped to the final dewatering process within the treatment works. Part way down the thickening zone a partial flow is recycled back to the reactor zone to act as the fresh ballast.

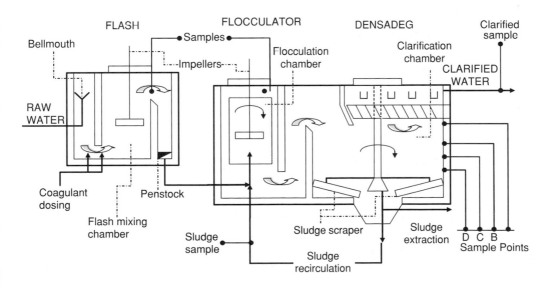

Fig. 4.12 Schematic of a Densadeg®.

In the final zone, clarification, the supernatant is passed upwards through a set of lamella plates to produce the final water quality.

Typical loading rates for the process are reported to be around 25 m hr^{-1} for drinking water application using coagulations, and as high as 120 m hr^{-1} for CSO applications. The key to the process is the requirement to generate the very dense flocs. Typical sludge recirculation rates are up to 10% of the flow at a concentration of between 10 and 20 g l^{-1} with anionic polymer dose rate of less than 0.5 mg l^{-1}. Control of the process comes from ramping up either component when necessary as in response to a storm event on a river.

The Densadeg® process like the Actiflo® process provides solutions with reduced hydraulic retention times (footprint), reduced coagulant demands and a rapid startup time. The latter is reported to be around 15–20 min enabling the processes to be used intermittently; this is of especial benefit for CSO applications but can be applied to storm events on drinking water plants and treatment of backwash effluent from filtration plants. Reductions in coagulant demand vary but are likely to be no more than a maximum of 50%. Most of the plants require the use of a polymer to generate sufficient floc strength to withstand the high shear rate zones encountered in the high rate systems. The other benefit is that the process is not as sensitive to coagulant dose such that periods of under-dosing do not cause such severe problems as with conventional systems.

Sirofloc®

The Sirofloc® process was developed in Australia and has been installed in the UK, Australia and the USA. The first plant in the northern hemisphere was installed at Redmires in 1988 with a capacity of 20 Ml d^{-1}.

The process works by contacting a 1–10 μm diameter powder of magnetic iron oxide (magnetite) with water at low pH. Under such conditions the magnetite acts as an adsorption medium removing any negatively charged material such as clay, algae and colloidal NOM. Following adsorption the powdered magnetite is agglomerated under the influence of a magnetic field which causes the powder to settle very quickly. The magnetite is cleaned by washing in a sodium hydroxide solution after which it can be reused. The process typically operates at a magnetite concentration of 1.5–2% by weight with the addition of a cationic polymer at a dose rate of around 1 mg l^{-1}. Settlement rates of up to 30 m hr^{-1} can be achieved but in practice the plants are operated at around 10 m hr^{-1}. Reported performance indicates typical removal rates of 99% turbidity, 96% colour and up to 99.9% on bacteria and viruses.

Table 4.2 Summary of clarifier design and removal.

Clarifier	Overflow rate ($m^3\ m^{-2}\ hr^{-1}$)	Retention time (min)	Turbidity removal efficiency (%)
Rectangular	1–2	120–180	90–95
Circular	1–3	60–120	90–95
Floc blanket	1–3	120–180	90–95
Lamella plates	5–12	60–120	90–95
Sand ballasted	<200	5–7	90–99
Sludge recirculation	<120	10–16	90–99
Magnetite	<30	15	90–99

4.4 Applications

In water treatment applications a properly designed and well-operated clarifier should produce clarified water of turbidity below 5 NTU representing removal efficiencies of around 90–95% (Table 4.2). The high rate systems can produce clearer effluents with residual turbidities close to 1–2 NTU in some reported applications.

References

Gregory, R., Zabel, T.F. & Edzwald, J.K. (1999) Sedimentation and Flotation. In: R. D. Letterman, ed. *Water Quality and Treatment*. McGraw-Hill Inc, New York.

Jarvis, P., Jefferson, B. & Parsons, S.A. (2004). Characterising natural organic matter flocs. *Water Sci. Technol. Water Supply.*, **4**, 79–87.

Plum, V., Dahl, C.P., Bentsen, L., Peterson, C.R., Napstjert, L. & Thomsen, N.B. (1998) The Actiflo method. *Wat. Sci. Technol.*, **37**(1), 269–275.

Dissolved Air Flotation 5

5.1 Introduction

Dissolved air flotation (DAF) is a process for the removal of particles/flocs (termed collectively as particles thus forth) by making the solid particles float rather than sink, as in clarification. The process works by attaching air bubbles to particles so as to reduce the density of the combined bubble–particle agglomerate. When the density is reduced to below that of water, the particles rise to the surface where sludge accumulates and can be removed by skimming. The more bubbles that are attached per particle, the faster the particle will rise and subsequently the higher the loading rate at which the process can be operated. The success of the process depends on two factors: (1) generation of enough bubbles and (2) conditions under which the bubbles and particles stick together.

A typical DAF plant, shown in Figure 5.1, is split into two key components. The flotation unit serves to separate the particle–bubble agglomerates from the water and the saturator dissolves air into water under an elevated pressure, typically 4–8 bar. Pressure saturated water is introduced into the flotation unit through a valve or nozzle that reduces the pressure very quickly. This causes the excess saturated gas to precipitate as tiny (30–80 μm) bubbles. The water is taken from the clarified outlet such that a partial recycle occurs. The ratio of flow from the saturator to the feed is a key operational variable, known as the recycle ratio (R). Immediately after the nozzle/reducing valve the saturator feed is mixed with the incoming feed water. The precipitated bubbles stick to the particles and the bubble–particle agglomerates rise to the water level and accumulate as a float. The difference in density between the float layer and the water lifts part of the float above the water level. Interstitial water from this part of

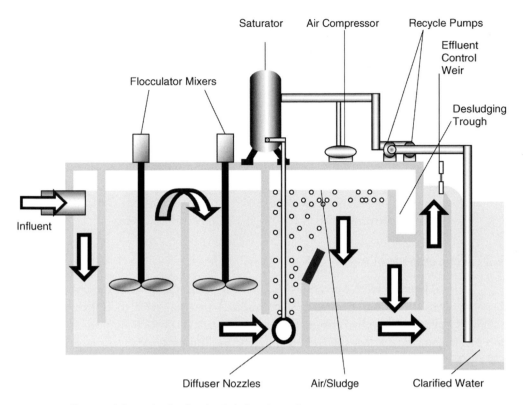

Fig. 5.1 Schematic of a dissolved air flotation unit.

the float drains therefore increasing the solids concentration, this is termed thickening. At pre-determined time intervals the float layer becomes too large and is removed by skimming or hydraulic lift.

Flotation was first used 2000 years ago, by the Greeks, to separate minerals from gangue. The technology has remained a crucial process in mineral processing ever since and has now been incorporated into a wide range of other industries. Initially DAF was used only when the particle density was less than that of water such that the unit was enhancing the natural process. In the 1960s the process started to be used for water and wastewater applications. Application in water treatment dates from the 1960s in South Africa and Scandinavia, but today the technology is becoming widespread in the UK (90 plants reported in 1990), North America (40 plants reported in 1990) and Australia (20 plants reported in 1997).

The majority of the plants were built post 1976 with the increasing popularity being attributed to the advantages of rapid start up, high loading rate, high solids capture and robustness to hydraulic and quality variation in the feed. These are offset in part by reported disadvantages of cost and complexity which restricted wide scale application of the technology in the 1980s and 1990s. Uptake of the technology has occurred most successfully in cases where the particles to be separated have low settling rates and so are unsuitable for sedimentation. The principal examples of this are in cases where natural organic matter or algae dominate the character of the water source.

This chapter will cover the process science and design of the DAF process as used in potable water production including a discussion of the key factors that influence operation and performance of the process. This will then be followed by a description of the main technology options including typical performance data for the main application areas.

5.2　Process science

DAF tanks have two zones (Figure 5.2). The purpose of the contact zone is to provide opportunities for bubbles and particles to collide and stick together, thereby forming the required bubble–particle agglomerate. In the separation zone, the bubble–particle agglomerates rise to the surface and are subsequently removed.

5.2.1　Contact zone

The contact zone can be viewed as either a heterogeneous flocculation process (Fukushi *et al.*, 1995) or by assuming that the rising bubble cloud

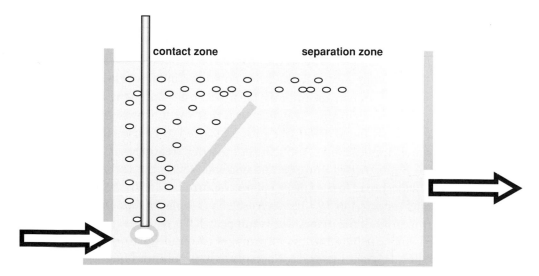

Fig. 5.2　Schematic of contact and separation zones.

acts like a filter bed such that the bubbles can be visualised as collector surfaces (Edzwald, 1995). Both approaches are reasonable but because of its direct link with deep bed filtration theory the collector surface approach will be described here.

The approach is based on the single collector collision efficiency concept used to describe particle transport to bubble surfaces and the exact details of the model can be found elsewhere (Edzwald, 1995). The approach has been successfully used for flotation processes and, more commonly, deep bed filtration. The model describes the transport of particles from the bulk water to a bubble surface according to the mechanisms of diffusion, interception and differential settling as described in the filtration chapter. The end expression relates the removal of particles to the total available surface area for contact and the residence time in the contact zone assuming plug flow conditions:

$$\frac{C_o}{C} = \exp\left(\frac{-3\alpha_{pb}\eta_T U_b \Phi_b t}{2d_b}\right)$$

where α_{pb} is the attachment efficiency, η_T is the single collision collector efficiency, U_b is the mean bubble rise velocity, Φ_b ($= c_b/\rho_{air}$) is the bubble volume concentration and t is the residence time in the contact zone. The model demonstrates that the efficiency of removal can be enhanced by

- Maximising the single bubble capture efficiency
- Maximising the attachment efficiency
- Increasing the bubble volume concentration
- Increasing the contact time
- Minimising the bubble size.

The model identifies three groups of variables that are crucial in the design and operation of DAF processes: α_{pb}, η_T and Φ_b/d_b. α_{pb} are variables related to pre-treatment and reflect the operation of the coagulation and flocculation stage of the process. α_{pb} depends on the coagulation conditions (type, dose, pH) and is empirically determined. Values for α_{pb} range from nearly 0 for poor attachment to 1 where all contacts lead to attachment. Typical values with good coagulation are between 0.1 and 0.5 meaning that an excess of bubbles is essential to ensure multiple contacts. Two conditions are necessary for good attachment: minimisation of the particle charge and a sufficiently hydrophobic surface.

The minimisation of particle charge is most commonly assessed through either zeta potential (ex situ) or streaming current (on line) measurement. The charge of the particles impacts on the attachment efficiency as strongly

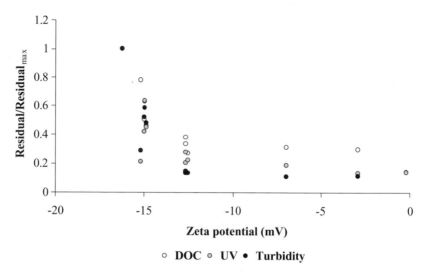

Fig. 5.3 Impact of zeta potential on removal in DAF.

charged surfaces repel like surfaces in the same way that two magnets with the same pole resist being brought together. As the charge on the particle's surface is reduced the barrier to contact between the bubble and the particle decreases and consequently α decreases. This continues until a certain threshold value, around -15 mV for DAF, below which no further improvement is observed as the barrier to contact has been removed (Figure 5.3).

The other requirement is that the surface of the particle exerts some degree of hydrophobicity. This is a measure of the affinity the surface has towards bonding with water and when this is strong (hydrophilic surface) the gas bubble cannot displace the water surrounding the particle's surface and hence no contact can occur. Surfaces are hydrophobic if they can only form van der Waals bonds with water and this is commonly achieved by organics with hydrocarbon tails such that surfactant and polymers are sometimes added to modify the surface properties of the particles to be removed. It is measured in terms of a contact angle made at the gas/solid, solid/liquid and gas/liquid interfaces. Angles greater than $25°$ are required for stable attachment but lower angles can be suitable if multiple contact points are made. Successful flotation requires both parameters to be favourable otherwise the bubbles and particles will bounce off one another when they collide and no removal will be achieved.

η_T is the probability of a collision between a particle and an individual bubble and is controlled by the combination of the particle and bubble sizes. In the case of the removal of particles above 1 μm and of low density the relationship is related to the square of the particle to bubble diameter ratio, $(d_p/d_b)^2$. Consequently, the probability of capture is maximised with small bubbles and large particles. Constraints on the energy required to produce small bubbles restricts this limit to around a mean bubble size of 40 μm (range 10–100 μm) and limitations on the rise velocity of the resultant flocs means that particle diameters are generally kept to around 100–400 μm. A minimum in the probability occurs around particle sizes of 0.5 μm as neither diffusion nor interception mechanisms operate well, therefore generating a decrease of an order of magnitude in the probability of collision compared to a 3 μm particle. Consequently, the process works best in the particle size range of 3–400 μm.

Φ_b is controlled by the air pressure and flow rate (recycle ratio) through the saturator and informs on the total available surface area for contact through changing the number of bubbles in the contact zone. Given that the pressure in the saturator generally stays constant once designed the main operating parameter is the recycle ratio. Assuming a saturator efficiency of 70% and a pressure of 4.8 bar, the bubble volume concentration increases from 2900 ppm to 8000 ppm as the recycle ratio is increased from 5 to 15% (Edzwald, 1995). This represents a bubble number concentration of between 9×10^4 and 2.4×10^5. Typical particle concentrations after coagulation and flocculation are of the order of 10^4–10^5 and so there are generally between 1 and 10 bubbles per particle. Sensitivity analysis of the model suggests the following:

1. The hydraulic residence time in the contact zone should be at least 90 s.
2. The contact zone should be plug flow.
3. The contact zone efficiency will exceed 95% under good attachment conditions and a bubble volume concentration of at least 5000 ppm.

5.2.2 Separation zone

Following successful collision and attachment, the bubble–particle agglomerate will rise to the surface. The impact of the attaching bubbles can be seen by using an adjusted Stokes' law expression:

$$v_{pb} = \frac{g(\rho_{water} - \rho_{pb})d_{pb}^2}{18\mu}$$

Table 5.1 Particle–bubble agglomerate properties.

		Number of bubbles	1	5	10	1	10
		Bubble size (µm)	40	40	40	100	100
	10	ρ_{pb} (kg m^{-3})	16.7	4.4	2.9	2.29	1.49
		d_{pb} (µm)	40.2	68.5	86.2	100	171
		v_{pb} (m hr^{-1})	3.1	9.2	14.5	19.6	57.3
	50	ρ_{pb} (kg m^{-3})	665.1	283.2	165.3	112.8	25.7
		d_{pb} (µm)	57.4	76.3	91.5	104	172.4
		v_{pb} (m hr^{-1})	2.2	8.2	13.7	18.8	56.8
Particle size (µm)	100	ρ_{pb} (kg m^{-3})	944.6	761.6	613.3	503.1	168.5
		d_{pb} (µm)	102.1	109.7	117.9	126	181.7
		v_{pb} (m hr^{-1})	1.1	5.6	10.6	15.5	53.9
	400	ρ_{pb} (kg m^{-3})	1004	1000	995.1	989.5	932.2
		d_{pb} (µm)	400.1	400.7	401.3	402.1	410.2
		v_{pb} (m hr^{-1})	NF	NF	1.6	3.3	22.4

$\rho_{air} = 1.29$ kg m^{-3}, $\rho_{water} = 1000$ kg m^{-3}, $\rho_p = 1005$ kg m^{-3}, NF = no flotation.

where v_{pb} is the rise velocity in m hr^{-1} and incorporates terms for the density and size of the new agglomerate:

$$\rho_{pb} = \frac{\left[\rho_p d_p^3 + B_n \left(\rho_b d_b^3\right)\right]}{\left[d_p^3 + B_n \left(d_b^3\right)\right]}$$

$$d_{pb} = \left[d_p^3 + B_n \left(d_b^3\right)\right]^{1/3}$$

Removal of the bubble–particle agglomerates occurs if the velocity of the agglomerate (v_{pb}) is greater than the overflow rate (v_L) of the separation zone (note not the whole flotation plant). Table 5.1 summarises the calculation for bubbles with a diameter between 40 and 100 µm contacting a particle with a size between 10 and 400 µm. The calculations show that one bubble can easily remove a particle of 50 µm but that multiple bubbles are required for particle of any larger size. Thus, it is advantageous to keep the particle size down to approximately 100 µm in order to obtain high rise velocities.

The most important feature of the calculations on the separation zone is that the rise velocity is not fixed and as such the loading rate of the process can be altered by changing the number of bubbles attached. In practice it is often difficult to assess the exact relationship between the concentration of bubbles injected and the average rise velocity of the bubble–particle agglomerates. Instead it is common to determine the relationship empirically, through pilot plant testing, between the amount of air injected and the maximum downflow rate at which the plant can operate. The most convenient means of assessing this is through the ratio of precipitated air to solids in the process, known as the air to solids ratio (a_s):

$$a_s = \frac{a_p Q_{sat}}{C_{feed} Q_{feed}} = \frac{(19.5P)R}{C_{feed}}$$

where P is the pressure in atmospheres, R is the recycle ratio (Q_{sat}/Q_{feed}), C_{feed} is the influent solids concentration and a_p is the mass of air precipitated per litre. At 20°C and above 3 bar equates to approximately 19.5 P. For any given a_s ratio and feed water there is a specific limiting downflow velocity, v_L, above which the bubble–particle agglomerates will get carried away with the effluent. The exact relationship is feed specific but normally takes the form of (Figure 5.4):

$$v_L = K_1 a_s^{K_2}$$

The air to solids ratio can be adjusted by changing either the recycle ratio or the pressure at which the saturator is running. In practice the saturator pressure is usually fixed once the plant is designed and so the a_s ratio is normally adjusted by changing the recycle ratio.

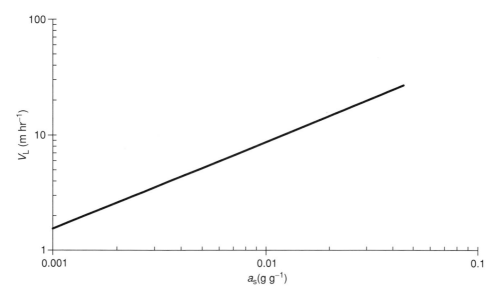

Fig. 5.4 Example relationship between a_s and v_L for algae.

5.3 Technology options

5.3.1 Traditional

Traditional DAF plants are either circular or rectangular. Circular tanks are normally used on small flotation plants, treating wastewater or sludge thickening applications, but some potable water plants exist with individual units of up to 20 m in diameter. An example is the Supracell® process which uses a rotating carriage to deliver feed water in the opposite direction such that the net velocity is zero enabling relatively shallow tank designs to be used. Rectangular tanks are the industry standard for large-scale applications due to advantages in terms of scale-up, simple design, easy feed introduction, easy float removal and compact design. Flotation tanks are typically designed with a depth of between 1.5 and 3 m at overflow rates of between 2 and 15 m^3 m^{-2} hr^{-1} (Table 5.2). Modern adaptations are enhancing the loading rates with 30 m^3 m^{-2} hr^{-1} being routinely achievable for some water types. Tanks often include a 60° baffle near the inlet to direct the bubble–particle agglomerates towards the surface and reduce velocity extremes in the incoming water (Figure 5.1). Individual tanks with footprints of 80 m^2 are in operation but the size is normally limited to practical issues of sludge removal. Smaller individual tanks are common which are banked together in series (Figure 5.5) providing the ability to desludge one tank while the others are still operating. The tanks must be covered or housed to avoid problems with wind, rain and potential freezing of the float layer during the winter.

Desludging is traditionally achieved mechanically whereby rubber or brush blades travel over the tank surface and push the float into a collection channel (Figure 5.6). An alternative is to use hydraulic desludging, or flooding, where the outlet to the tank is closed and the water level increases due to the incoming flow. Hydraulic systems are especially good for fragile flocs as they generate less agitation, but generally result in higher wastage rates (up to 2%) and relatively low float solids concentrations.

Table 5.2 Summary of DAF design data (adapted from Edzwald, 1995).

Parameter	Netherlands	Finland	UK	USA	South Africa
Loading rate (m hr^{-1})	10–20	3–8	5–12	6–10	
% recycle	6.5–15	5.6–42	6–10	8–40	6–10
Pressure (kPa)	400–800	300–700	310–834	480–620	400–600
Method of desludging			Mechanical scrapper	Mechanical scrapper	
Frequency (1/hr)		0–24	0–2	Continuous	
Sludge % solids	0.1–8	3	0.3–3	<1	

Fig. 5.5 Rectangular flotation tanks.

5.3.2 Combined flotation and filtration

Conventional DAF has a loading rate very similar to deep bed filtration processes (see Chapter 6), which has enabled the two processes to combine to provide a low footprint high product water quality process. The technology was first pioneered in Sweden in the 1960s but its use has

Fig. 5.6 Mechanical desludging.

been reported from many other countries since. A rapid gravity sand or anthracite–sand filter is incorporated in the lower section of the flotation tank such that the effluent from the flotation stage flows directly onto the filter bed. Consequently, the combined unit requires additional depth, up to 2 m, to house the filter bed and underdrain system. As the filter bed acts as a batch process (see Chapter 6) the flotation process must be intermittently stopped to permit backwashing of the filter. The process offers two main advantages over traditional systems: (1) disturbances in the flow which may cause the flocs to break are minimised and (2) the system can easily be switched to filtration-only operation with the DAF coming on line only when the water becomes excessively polluted. An example of this is during the treatment of reservoir waters experiencing seasonal algal blooms and requiring pre-treatment before filtration to avoid excessive head loss build up during periods of heavy algal growth.

5.3.3   Counter current flotation (CoCoDAFF®)

Traditional DAF plants work in a co-current mode whereby the bubbles and incoming water are contacted together at the bottom of the tank and then the combined flow travels in a horizontal direction in the separation zone. CoCoDAFF® operates in a counter current direction with the flow maintained in a vertical direction enhancing both bubble–particle interactions and the time in the contact zone (Figure 5.7). A key attribute of the system is that the incoming flow is evenly distributed across the flotation tank through a series of submerged laterals and distribution cones (Figure 5.8). By introducing the bubble cloud counter currently to the flocs, the majority of particles have already been captured avoiding breakage problems around the high shear rate areas where the bubble cloud is introduced.

The CoCoDAFF® process was originally developed to overcome operational problems with seasonal algal blooms which were blocking the existing sand filters. The process offers two key advantages over traditional DAF plants: (1) the potential for floc damage by the injection of the recycle flow is greatly minimised and (2) its operational flexibility, such that the DAF component is only turned on during periods of operational difficulty.

5.3.4   Combined flotation and lamella plates (DAFRapide®)

DAFRapide® is a combination of DAF with lamella clarification that enables greatly enhanced loading rates of greater than 40 m^3 m^{-2} hr^{-1} and

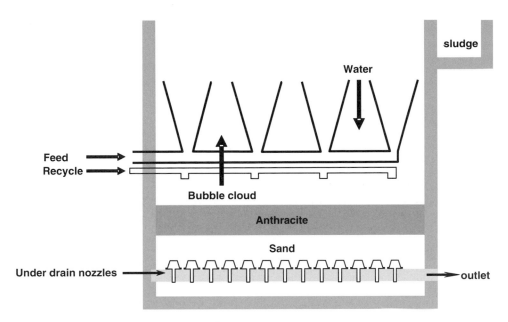

Fig. 5.7   Schematic of CoCoDAFF®.

Fig. 5.8   CoCoDAFF® distribution nozzles (with kind permission from Thames Water).

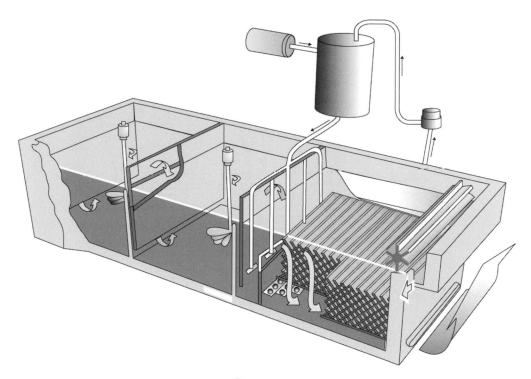

Fig. 5.9   Schematic of DAFRapide® (with kind permission from Purac Ltd.).

reduced flocculation times of 10–15 min (Figure 5.9). The lamella plates function in the opposite way to a traditional clarifier in that the water leaves the bottom of the plates. As the loading rate is increased, the bubbles cloud expands downwards towards the outlet and ultimately penetrates between the lamella plates. Small bubble–particle clusters and free bubbles collect on the surface of the plates until they grow to a sufficient size that the buoyancy force lifts them off the plate and they float to the surface. This action enables much greater loading rates as cluster agglomerates tend to have very high rise velocities. The process has been initially applied to wastewater applications but plants are currently being constructed for potable applications.

5.4   Applications

The majority of applications of DAF processes in potable water treatment are in situations where low density particles with slow settling rates are encountered which restricts the effectiveness of clarification processes. Consequently, the principal application areas are upland sources (low

Table 5.3 Summary of DAF performance.

Source	Process	Loading rate (m hr^{-1})	R(%)	Turbidity	Removal (%) [a]DOC/[b]colour	Algae
Upland	Traditional	6.3	8	94	[a]90	–
Upland*	Traditional	8.5	7–10	–	–	98.6
Upland	Traditional	16	–	82	64	–
Lowland	Traditional	6	8	–	–	89.9
Lowland	Traditional	5.2	11	–	–	54.1
Lowland	CoCoDAFF®	10	10	90	–	–
Lowland	CoCoDAFF®	9.2	–	96	[b]97.5	98
Lowland	DAFF	8–12	5–12	96.5	[b]98	→100
River	DAFRapide®	15	7	90[†]	[a]84 (UV)	–

*Largest DAF plant in the world in 1997.
[†]Target level.

alkalinity, coloured) and lowland reservoirs (algae-rich waters). Less common, but still regularly reported, are applications for lowland, mineral-bearing (high alkalinity) river waters and low turbidity, low colour waters (Gregory *et al.*, 1999). To illustrate, in the case of algae removal, DAF reduced the concentration of *microcystis* from 102 000 cells ml^{-1} to 2000 cells ml^{-1} (98%) compared to 24 000 cells ml^{-1} (76%) with sedimentation.

Removal efficiencies of 90% and above are common for turbidity, natural organic matter (measured as DOC or UV$_{254}$) and algae removal (Table 5.3). In the case of algal and zooplankton (*Daphnidae, Cyclopoida*) performance is species dependant with free swimming species generally being harder to remove, with removal efficiencies down to 40–50% in some case studies.

References

Edzwald, J.K. (1995) Principles and applications of dissolved air flotation. *Wat. Sci. Technol.*, **31**(3–4), 1–23.

Fukushi, K., Tambo, N. & Matsui, Y. (1995) A kinetic model for dissolved air flotation in water and wastewater treatment. *Wat. Sci. Technol.*, **31**(3–4), 37–47.

Gregory, R., Zabel, T.F. & Edzwald, J.K. (1999) Sedimentation and Flotation. In: R.D. Letterman, ed. *Water Quality and Treatment*, McGraw-Hill Inc, New York.

Filtration Processes 6

6.1 Introduction

Filtration processes are used principally for the removal of particulate material in water including clays and silts, micro-organisms and precipitates of organics and metal ions. The process of filtration involves passing water, containing some physical impurity, through a granular bed of media at a relatively slow velocity. The media retains most of the contaminants whilst allowing the water to flow. The particles that are removed are typically much smaller (0.1–50 μm) than the size of the filter media (500–2000 μm) such that virtually no simple straining occurs and removal is based on particles colliding and sticking to filter grains as the water flows past.

Filtration through granular media, particularly sand, is one of the oldest and most widely used water treatment processes. Early reported applications of the concept include the Romans in 150 AD and the Italians in 1685. However, its first reported use as a municipal sand filter dates back to 1804 when a slow sand filter was installed in Paisley in the UK. In 1827, Robert Thom was granted a patent for a filter with a false floor and cleaning by backwash, and in 1829 the engineer James Simpson installed a similar design, which required scrapping rather than backwashing of the so called *English filter*, in Chelsea, UK. The immediate benefits in terms of control of waterborne diseases such as cholera and typhoid led to the wide-scale uptake of the technology across the UK. In the late half of the nineteenth century the need for large quantities of safe drinking water in the USA led to the development of coarser, more rapid filters. The first filters were designed by trial and error and although Darcy published his work in 1856 it was not until 1937 that the Kozeny–Carman equation, which describes flow through the bed, first appeared. Consequently, filters were designed using 'rules of thumb' developed from the early empirical designs and, although research into granular bed filters continues today,

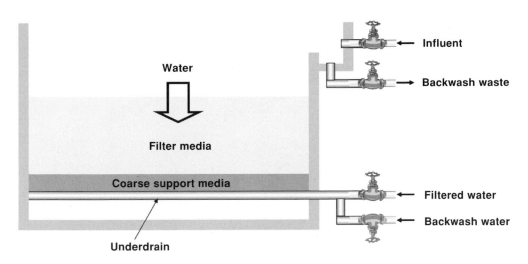

Fig. 6.1 Schematic of a deep bed filter.

those rules of thumb have stood the test of time and are still widely used on a day-to-day basis.

A typical filter consists of a rectangular or square tank containing the media, with a suitable arrangement of pipework to collect the filtered water and when necessary provide the water/air for the backwash (Figure 6.1). The size of individual filters is restricted by the need to distribute the flows evenly such that they are usually less than 10 m by 10 m although filters up to 150 m² exist. To provide the required surface area filters are usually banked together in a series of parallel tanks. The advantage of this is that individual tanks can be backwashed without a loss of service to the whole works.

The media sits on top of an under-drain system which supports the media, collects the filtered water and delivers the backwash flows. A wide range of systems are available including the earliest types which are the *manifold lateral systems*, where perforated pipes are located at intervals along a manifold. More modern systems include the *fabricated self-supporting system* and *false floor under-drain with nozzles*. In fabricated systems, blocks are constructed out of polymers such as HDPE which either support traditional gravel layers or include purpose-built media retention plates, containing opening sizes between 300 and 500 μm, on which the sand media can directly sit (Figure 6.2).

In false floor systems, a concrete slab is located 0.3–0.6 m above the filter floor to provide an under-drain plenum. Nozzles are inserted at regular intervals of about 10–20 cm across the slab (Figure 6.3). The nozzle stems protrude into the plenum and may have one to two small holes drilled into

Fig. 6.2   A fabricated filter floor block (with kind permission from US filters).

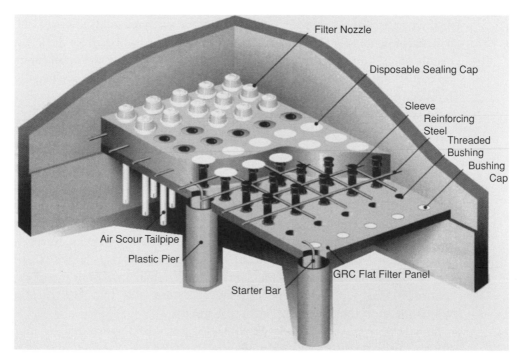

Fig. 6.3   A schematic of a false filter floor (with kind permission from AMT Systems Ltd.).

them to provide better air distribution across the bed. The nozzle openings are generally larger than the sand such that graded gravel layers are placed over the nozzle to prevent sand passing through.

During the operation of the process particles accumulate within the void spaces of the filter bed, decreasing the effective pore size and hence increasing the resistance to flow. The result is either an increase in the pressure drop (constant flow operation) or a decrease in flow (constant pressure operation). As the filtration process continues, the top layers of the bed become laden with deposits such that the grains cease to retain any additional particles. Consequently, the operation of the bed moves in a plug as progressive layers of the bed become exhausted until the bottom of the bed is reached where no capture occurs at all. When either particle breakthrough or the head loss becomes too large, the filter requires cleaning by back flowing water, in conjunction with air, to dislodge the particles retained and so clean the filter media. Hence, the filtration process is a batch operation that typically lasts between 12 and 36 hr, known as the filter cycle, which can be divided into seven distinct zones (Figure 6.4):

1. *Lag phase*, where clean backwash water passes out of the filter from the under-drains.
2. *Pre-ripening phase*, where the effluent quality becomes poorer, caused by dirty backwash water remnants within the bed.
3. *Second pre-ripening phase*, where the effluent quality further deteriorates because of dirty backwash water remnants above the media.

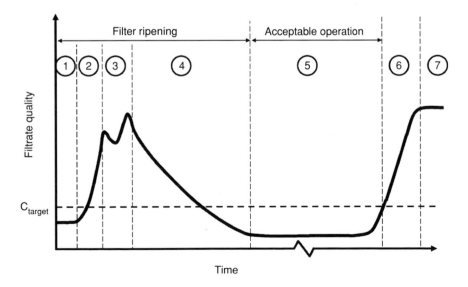

Fig. 6.4 A typical filter cycle.

4. *Ripening phase*, where the filtrate quality starts to improve to its steady-state value.
5. *Effective filtration*, where the filter is operating at its optimum level and represents the main period of operation within the cycle.
6. *Breakthrough phase*, where the filter quality begins to deteriorate as the filter reaches its capacity.
7. *Spent phase*, where the filter effluent concentration has reached a maximum and wormhole flow may be occurring.

The most complex part of the cycle is through the initial period and a lot of attention is placed on reducing the ripening period and the extent of the two turbidity spikes during stages 2 and 3 of which filtering to waste and the use of coagulants are the most common methods. The lag and deterioration phases (1 and 6) are not often reported and typically take about 10 min each. The two most important operational issues with regard to the filter cycle are intermittent operation and changes in the filtration rate as both cause hydraulic shocking through the bed and dislodge large quantities of particles causing turbidity spikes and so should be avoided, if at all possible. However, it is often not possible and then the key is to minimise any potential problems by gradually increasing the rate rather than rapid changes.

This chapter will cover the process science and design of the filtration processes as used in potable water production including a discussion of the key factors that influence operation and performance of the process. The chapter will then describe the main technology options including typical performance data for the main application areas.

6.2   Process science

6.2.1   Removal

The capture of particles within the filter bed involves a two-step process of collision and attachment. In the first step particles are brought close to the surface of the sand grains by a range of different mechanisms described below. Once there, the balance of surface forces between the two surfaces determines whether attachment will occur. The scale of the two steps is several orders of magnitude different such that they can be considered separately.

Collision
Flow through the filter is laminar over the loading rates typically observed such that particles are carried along flow streamlines unless a transport mechanism causes them to deviate. Considerable work has been conducted

on understanding the possible forces that cause the required collisions, and in most practical cases three mechanisms tend to dominate. Expressions for the probability of capture based on each individual mechanism have been adapted from the original air filtration expressions.

(1) Diffusion

All particles will be moved in a random pattern away from their streamline due to the bombardment of surrounding molecules, known as Brownian motion. It is generally accepted that particle diffusion is only applicable for particles of less than 1 μm in diameter:

$$\eta_D = 0.9 \left(\frac{KT}{\mu d_p du} \right)^{\frac{2}{3}}$$

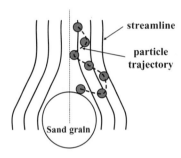

(2) Interception

If the particle radius is greater than the distance between the flow streamline and the grain, the particle will contact the grain as the fluid flows past whilst remaining on the streamline:

$$\eta_I = \frac{3}{2} \left(\frac{d_p}{d} \right)^2$$

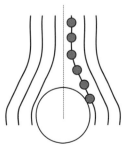

(3) Sedimentation

When the fluid flow is directed downwards sedimentation effects will cause the particles to settle vertically onto the grains. In this way the individual filter media are acting like very small sedimentation tanks. The probability of capture is characterised by the dimensionless ratio of the Stokes settling velocity and the approach velocity of the fluid:

$$\eta_S = \frac{d_p{}^2(\rho_P - \rho)g}{18\ \mu u}$$

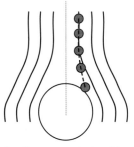

Where d_p is the particle diameter (m), d is the diameter of the filter media (m), u is the filtration velocity (m^3 m^{-2} hr^{-1}), m is the viscosity (kg m^{-1} s^{-1}), K is Boltzman's constant and T is the absolute temperature (K).

The total probability of capture is calculated by summing the probabilities of the individual mechanisms which shows that the process is dominated by diffusion for very small particles (<1 μm) and interception or sedimentation for large particles (>1 μm). A minimum occurs at 1 μm where the crossover from diffusion control to sedimentation or interception control occurs (Figure 6.5). Thus, at first it may seem strange but very small particles (sub-micron) are more efficiently removed than

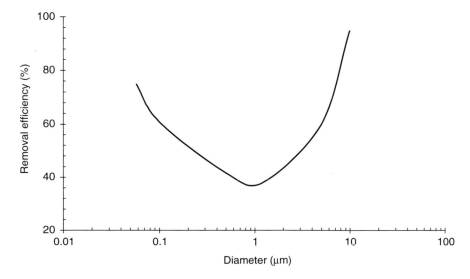

Fig. 6.5   Removal efficiency as a function of particle size.

larger particles (micron). Analysis of these expressions suggests that the probability of capture will be reduced by high filtration rates ($u^{-1}, u^{-2/3}$), large media diameters ($d^{-2}, d^{-2/3}$), colder temperatures ($\mu^{-1}, \mu^{-2/3}$) and smaller light particles ($d_p^2, (\rho_p - \rho)$) for post 1 μm particles.

Attachment

When a particle approaches a grain to closer than 100 nm, short-range forces begin to become important. There are several types of interactions which may be critical, the best known of which are van der Waals attractions and electrical repulsions. Together these make the basis of the classical DVLO theory of colloid stability, which calculates the total interaction force between two surfaces as they approach one another (see Chapter 3).

When the total interaction force is high, a barrier will exist to contact such that the particles bounce off the sand grain as they collide. When the interaction force is reduced, typically by reducing the surface charge of the particle, the barrier is removed and contact can occur. Particle charge is most commonly assessed through either zeta potential (ex situ) or streaming current (on line) measurement. In reality, a threshold value of charge is reached below which there is no further improvement as the barrier to contact has been minimised. In deep bed filtration processes this threshold value is normally around −15 mV (Figure 6.6), although the exact value varies according to the system under investigation. Comparison of all the reported systems suggested that as long as a particle's charge is less than −10 mV, no barrier to removal should exist due to attachment. The differences observed in Figure 6.6 between filled and open circles represent the performance of two filters with different media sizes. Thus, the zeta potential is a good indicator that the process is working optimally but does not inform about what the likely optimum turbidity will be.

Filtration descriptions

Incorporation of the mechanistic expressions into an analysis of the trajectories of the streamline enables the performance of the filter to be predicted (Elimelech *et al.*, 1995):

$$\ln\left(\frac{C}{C_0}\right) = -\frac{3}{2}\frac{(1-\varepsilon)\alpha\eta_0 L}{d}$$

where α is the attachment efficiency, ε is the bed porosity, η_0 is the collision efficiency and L is the bed depth (m). The above approach can be used to investigate the likely impact of filtration variables on performance (Figure 6.7). The first two plots, (a) and (b), show the effect of altering

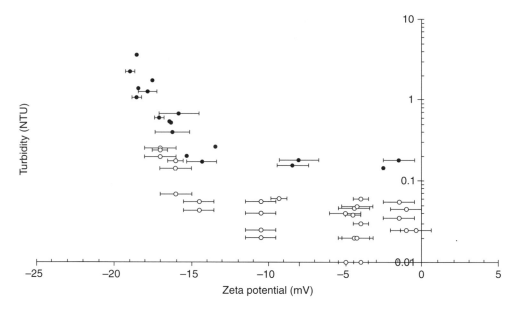

Fig. 6.6 Relationship between zeta potential and filtration performance (with kind permission from United Utilities).

the attachment efficiency from poor or no chemical addition ($\alpha = 0.05$) to optimal conditions ($\alpha = 1$). Plots (a) and (b) demonstrate that a media depth of 60 cm can effectively remove particles smaller than 100 nm (viruses) and particles larger than 7 µm (protozoan cysts). The limitation is seen around 1 µm (the size of pathogenic bacteria) where the efficiency is at a minimum. Filtration rates (c) most seriously affect the removal of smaller particles as particles larger than 1 µm are removed principally by interception which is independent of velocity. Grain size (d) affects the removal of particles across the entire size range and is seen as the major influence on filtration performance.

The alternative approach to describing filtration performance is described as *phenomenological* modelling as it makes no attempt to consider the mechanisms of particle removal. Instead the approach considers the filter from a macroscopic point of view and describes the performance by conducting a mass balance over the filter in conjunction with an empirical rate expression. The mass balance describes the suspended solids' gradient with the rate of change of specific deposit within the filter; whereas, the rate expression describes the rate of change of particles as a function of their concentration. Together they provide a description of how the filter operates in relation to both depth and time. Solution of the above expressions is rather complex and generally needs a computer simulation to yield

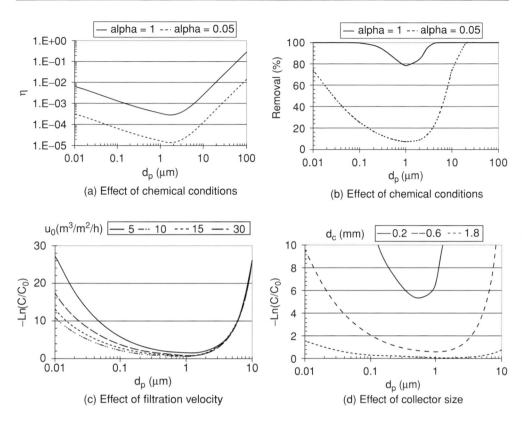

Fig. 6.7 Filter performance according to fundamental (microscopic) models. $U = 0.14$ cm s^{-1}, $L = 60$ cm, $\varepsilon = 0.4$, $dg = 0.5$ mm, $T = 293$ K, $\rho_p = 1.05$ g.cm^{-3} (n.b. numbers typical of water treatment).

useful predictions.

$$\text{Mass balance:} \quad -U_0 \frac{\partial C}{\partial y} = \frac{\partial \sigma}{\partial t}$$

$$\text{Rate expression:} \quad \frac{\partial C}{\partial y} = -\lambda C$$

where σ is the specific deposit (volume of deposit per filter volume), λ is the filter coefficient and represents the key descriptor of any individual filter's ability to remove particles, y is the filter depth and t is the run time. Comparing the two approaches:

$$\lambda_0 = \frac{3}{2} \left(\frac{1 - \varepsilon}{d} \right) \eta_0$$

As the bed becomes laden with captured particles it becomes less efficient at removal as the streamlines straighten out and the local velocity within the pores increases generating more shear forces. As this process continues through the bed, the effluent concentration starts to increase and the filter must be taken out of service and backwashed (Figure 6.4). In practice, the models only work effectively during the initial stages of filtrations and attempts to predict filter run times are largely inconsistent. Instead the capacity of a filter can be estimated from an empirical expression but ultimately pilot trials will be required for any specific water. Typical capacity values are around 1000 g m^{-3} of media for sand and 2000 g m^{-3} for anthracite.

6.2.2 Hydraulics

The head loss (i.e. pressure drop) that occurs as water passes through a clean filter bed can be calculated from the well-known Kozeny–Carman equation:

$$\frac{\Delta P}{L} = \mu \left(\frac{K (1 - \varepsilon)^2 S_v^2}{\varepsilon^3} \right) \frac{dV}{dt} \frac{1}{A}$$

where K is the Kozeny constant which is 5 for a fixed bed or slowly moving bed and 3.4 for a rapidly moving bed, dV/dt is the volumetric throughput (flow rate if constant) and S_v is the specific surface area of the media. Within the typical ranges used in water treatment a simplified expression can be used as the pressure drop increases in an approximately linear fashion with velocity:

$$\frac{h}{L} = 1.02^{(15-T)} k_h u$$

where k_h is a head loss coefficient specific for a particular media (size and type) (Table 6.1).

During the operation of the filter, the bed becomes laden with retained particles. The increase in pressure drop is proportional to the amount of

Table 6.1 Typical values for the head loss coefficient.

Media	Mesh	Size (mm)	k_h
Sand	16/30	0.5–1.0	0.06
Sand	14/25	0.6–1.2	0.05
Sand	8/16	1.0–2.0	0.04
Anthracite	–	1.2–2.5	0.01
Garnet	25/52	0.3–0.8	0.12

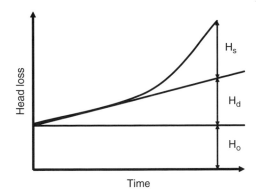

Fig. 6.8 Idealised head loss development profile.

material collected and so the total head loss in the filter becomes:

$$H = H_0 + K_v C_0 t$$

where K_v is a constant, H_0 is the initial head loss, H_d is the head loss due to operation (Figure 6.8). The steeply increasing exponential curve of head loss is due to the formation of a surface mat of deposit on the inlet face of the filter bed. This mat indicates straining is taking place – the very antithesis of depth filtration. The consequence of this is a very much shorter filter run with little or no utilisation of the bed depth taking place. To reduce these effects any of the following can be tried:

- Reduce inlet concentration
- Remove larger particles
- Replace inlet filter media with a coarser grain size
- Increase velocity.

6.2.3 Backwashing

The goal of backwashing is to keep the filter acceptably clean through a combination of fluid shear forces, grain collisions and abrasion between grains. Filter water is back-pumped up through the filter bed at a sufficient rate to fluidise the bed and expand it between 5 and 30%. At the point when the bed starts to expand the lift force generated by the upflowing water matches the weight force of the media. A force balance at this point enables the minimum fluidisation velocity (v_{mf}) to be calculated:

$$v_{mf} = \frac{g}{180} \frac{(\rho_s - \rho)}{\mu} \frac{\varepsilon^3 d^2}{(1 - \varepsilon)}$$

Table 6.2 Fluidisation velocities for different media (adapted from Droste, 1997).

Mean size (mm)	Flow rate required for a 10% bed expansion at 25°C (m hr⁻¹)		
	Anthracite (s.g. = 1.7)	Sand (s.g. = 2.65)	Garnet (s.g. = 4.1)
2.59	90.4	–	–
1.84	58.7	100.2	–
1.30	38.4	66.0	119.8
0.92	24.2	40.1	78.2
0.65	17.1	22.0	53.8
0.46	–	13.2	33.5
0.27	–	–	15.4

Examination of the above equation reveals that the following variables are important in determining the fluidisation velocities of filter beds:

- Media grain size
- Media density
- Media voidage and packing
- Temperature (as it affects the water viscosity).

As the backwash velocity increases up to the minimum fluidisation velocity the pressure drop across the filter holds to Darcy's law as the grain is still fixed. Beyond the v_{mf}, the bed starts to expand and the head loss remains constant as all the energy is used to maintain the grains in suspension. The degree of bed expansion at the point of fluidisation is calculated by

$$\frac{l_e}{l} = \frac{1 - \varepsilon}{1 - \left(\dfrac{v}{v_t}\right)^{0.2}}$$

Where l is the bed depth and l_e is the bed depth after expansion. Typical upflow velocities to achieve a 10% bed expansion are given in Table 6.2. It is important to remember that the fluidisation velocity is a function of both the density and viscosity of water and that these parameters vary with temperature. For example to achieve the same bed expansion at 5°C as at 15°C requires an increase in flow rate of about 50%.

6.3   Technology options

6.3.1   Rapid granular bed filtration

All filters consist of tanks containing media with a suitable arrangement of pipework to collect filtered water and allow backwashing. The various types of filters can be classified according to

1. Pressure or gravity flow
2. Type of media
3. Method of flow control
4. Backwash method.

Pressure or gravity flow

Rapid granular filters can be operated under gravity or pressure. In gravity filters, relatively shallow beds, 600 mm, are used to reduce pressure drops across the filter. The majority of plants are square or rectangular open tanks made from concrete. Typically loading rates depend on the application but range between 10 and 15 m hr^{-1} and are used for lowland river and more recently upland sources.

In pressure filters the filter shell is a pressure vessel of either carbon steel, glass fibre reinforced plastic (GRP) or stainless steel and may be of either *vertical* or *horizontal* design. Vertical filters have been constructed up to about 4.5 m diameter but larger surface areas can be accommodated in 3 m diameter horizontal types. In this design the top of the bed is usually situated just on or above the centre line of the vessel shell and the effective bed cross-sectional area is approximately the diameter of the shell multiplied by the length of the shell. Typical loading rates are the same as gravity filters although larger media are sometimes used which permit higher rates. Pressure filters are mainly employed for small works, good quality soft upland water treatment and groundwater (both for iron and manganese removal). The latter has the advantage that the water does not require re-pumping to transfer it to supply.

Types of media

The most common types of media used in granular filters are sand, crushed anthracite and garnet (Table 6.3). Traditionally all beds were single media, but problems were observed in that after backwashing the media stratified

Table 6.3 Properties of different media used in filtration (adapted from Cleasby & Logsdon, 1999).

Type	Position in bed	Media	Depth (m)	Specific gravity	Size (mm)
Mono-media	All	Sand	0.6–1.0	2.6	0.5–1.0
Dual media	Top	Anthracite	0.3–0.4	1.5	1.2–2.5
	Bottom	Sand	0.3–0.6	2.6	0.5–1.0
Multi-media	Top	Anthracite	0.4–0.5	1.5	1.2–2.5
	Middle	Sand	0.3–0.4	2.6	0.5–1.0
	Bottom	Garnet	0.1–0.2	4.2	0.2–0.3
Coarse monomedia	All	Sand	1–3	2.6	1–2

into size grades with the larger grains at the bottom and the smaller grains at the top. The impact of this is to produce filters with high efficiency top layers with a correspondingly rapid pressure loss development such that filter run times are reduced. The most common method of alleviating this problem is to use multi-media beds, typically of a coarse anthracite above sand where the upper anthracite layer acts to remove the bulk of the solids and the sand layers polish. This can be taken further by using a layer below the sand of fine garnet, a system which is gaining popularity in the US. The difference in the specific gravity of the materials means that the larger particles will still settle more slowly after fluidisation such that the bed remains segregated in layers of different material.

Alternative approaches involve using either upflow filtration or coarse deep bed filters which are 2–3 times the depth of a normal filter. Both technologies are commonly used as a tertiary treatment step in wastewater treatment where the suspended solid loads onto the beds are much higher, but are relatively uncommon in potable applications. The exception is in the case of coarse deep beds, which are gaining in popularity for direct filtration, where the solids concentration is similar to wastewater applications. In some applications of taste and odour problems the sand media is replaced by activated carbon of a similar size which provides a combined process of adsorption and filtration. Reported experience suggests the carbon beds can last up to 5 years before regeneration, and that a small sand layer may be required below the activated carbon to provide a final polish where low turbidities are required.

Comparison of filters with different-sized media is usually based on a concept of equivalent depth whereby the ratio of bed depth to grain diameter (L/d_e), where d_e is the effective size of the media, is kept constant at around 1000 for mono- and dual beds, and 1250 for triple beds and coarse mono-media beds. In the case of multi-media beds the L/d_e ratio is the sum of the individual layers.

Method of flow control

Normal practice is to provide a minimum of three or four filters and preferably six or more to generate sufficient flexibility to adjust to periods when an individual filter is out of service for repair or being backwashed. There are a number of flow control systems that can be used which are grouped as either equal (also known as constant) rate or declining rate systems. Constant rate systems are strictly only possible when the incoming flow is constant, so it is better to refer to them as equal rate.

Equal rate filtration is the most common method and maintains the flow through each bed whilst allowing the head loss to increase as the bed becomes laden. The most common method of equal rate control is called *proportional level control* (Figure 6.9(a)) which uses a flow controller in the filtrate pipe to maintain the flow at a preset rate. Influent water is split equally between the operating filters with a single transmitter controlling the water level within the influent chamber within pre-set bands. Water enters below the water level in the filter and the outlet rate is matched to the incoming flow by the controller. The main advantage is that each filter

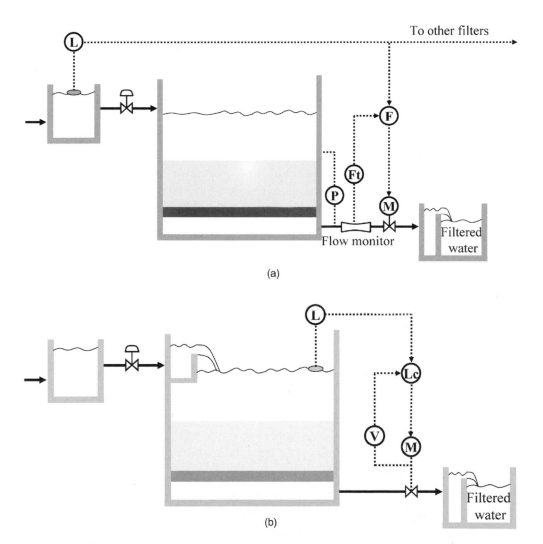

Fig. 6.9 Equal rate control systems: (a) proportional level control and (b) proportional level flow splitting. F = flow controller, Ft = flow transmitter, L = level transmitter, Lc = level controller, M = motorised valve, P = pressure transmitter, V = valve position transmitter (adapted from Cleasby & Logsdon, 1999).

in a battery operates at a constant flow and is not subject to surging when one of the other filters is taken off line for washing.

Alternative equal rate systems exist which allow variable levels within the individual filters. This is achieved by splitting the flow through an inlet weir box on each filter above the water level in the bed (*variable level flow splitting*). The level in each bed is different and not balanced but such systems are simple and require no real instrumentation. More sophisticated measures can be employed by using level control transmitters to control the outlet valve such that the opening is proportional to the water level in each bed (*Proportional level flow splitting*: Figure 6.9(b)). The benefit of such systems is that transient conditions are smoothed out but at the cost of an increased filter cell height and a variable water volume held in the filter to be treated, which may impact on floc properties.

Declining rate control relates to maintaining a consistent level in all filters, through a common inlet head, whilst allowing the rate to decrease as the filter bed becomes laden. Flow enters below the normal water level in the bed. It is common to use some form of flow restriction to limit the initial flow rate after backwash to a maximum of 130% of the design flow. This system of operation works well when the total flow is near the design level but under conditions of low flow, say 50% of design, the cleanest filters take the highest proportion of the flow leading towards a unification of when the beds require backwashing. Declining rate control is no longer acceptable in the UK and is being discouraged in the US.

Backwash

In modern filtration plants the backwash cycle is triggered by either a pre-determined level of head loss, an increase in the filter effluent turbidity or a preset time interval. The latter is most common such that filters backwash every 24 hr or so. A number of different techniques are employed:

- Fluidising water wash
- Fluidising water wash plus surface jets
- Air scour followed by a fluidising water wash
- Simultaneous air and water wash followed by a fluidising water rinse.

Backwashing with water is the traditional approach such that effluent is back-pumped through the under-drain system. Initially the inlet valve is closed and the water level allowed to drop to the backwash weir height. The backwash system should be turned on gradually over a 30-second period to avoid unnecessary disturbance to the media or sudden pressure surges on the under-drain system. Fluidisation is continued until the wash water is reasonably clear, typically around 10 NTU, at which point the filter is put back into service. Backwash volumes of between 3–7 bed volumes are common representing around 2–5% of the filter's throughput delivered at a rate of between 20–30 m³ m⁻² hr⁻¹ for a standard sand filter.

Water-only backwashes are not always very effective such that it is common to use an auxiliary scouring system to enhance the backwash. In the US, filters that were installed before 1980 sometimes use surface wash systems by injecting jets of water just above the filter bed from 1 to 2 min before the backwash until 2–3 min before it stops. The orifice jets are typically 2–3 mm in diameter, angles at 15–45° and run at 5–10 m hr⁻¹ at a pressure of 350–520 kPa. In many cases, an air scour is preferred as it assists through the whole depth of the bed and covers the full area. The air can be applied before or simultaneously with the backwash, and can be supplied through either a dedicated system or as part of the under-drain arrangement.

Traditionally in the UK, air is applied before the backwash to dislodge particles ready to be flushed away. Typical rates are of the order of 20–30 Nm³ m⁻² hr⁻¹ for 5 min, although rates as high as 70 Nm³ m⁻² hr⁻¹ have been reported (Table 6.4). The use of air scour in this way causes a lot of agitation in the top layers of the bed but not very much in its depth, and as such is most effective if the bed is not too heavily loaded in its bottom layers. In Europe, and increasingly in the UK, preference is for a combined air and water wash, where the bed is not fluidised. Instead, a process called collapse-pulsing occurs which relates to air pockets in the bed but has been shown to truly reflect the onset of three-phase fluidisation. Typical air flow rates of 50–90 Nm³ m⁻² hr⁻¹ are used in combination with a water flow rate of 10–20 m³ m⁻² hr⁻¹ (Table 6.4).

Most of the operational problems associated with filters are due to one of the aspects of the backwash system. The most serious problems occur due to under-drain failures which are generally associated with bad

Table 6.4 Typical backwash conditions (adapted from Cleasby & Logsdon, 1999).

Filter medium	d_{10} (mm)	Sequence	Air rate (m hr⁻¹)	Water rate (m hr⁻¹)
Fine sand	0.5	Air then water	35–55	30–40
Fine anthracite	1.0	Air then water	55–70	35–50
Coarse sand	1.0	Air + water then water	55–70	15–20 15–40
Coarse sand	2.0	Air + water then water	110–150	25–30 25–60
Coarse anthracite	1.5	Air + water then water	55–90	20–25 20–50

design or installation causing nozzle blocking or bed collapse. Typical problems are associated with blocking from debris left in the filter during construction, solids in the back wash, where steel is used and allowed to rust, and biological growth.

The other set of problems is associated with insufficient backwash, generally in relation to the auxiliary system. The most commonly reported events are filter cracks, mudballs and filter boil. Filter cracks are generated by the backwash leaving a thin compressible layer of material on the grains. As the pressure drop increases during subsequent cycles the grains are squeezed together and cracks start to appear permitting short-circuiting through the filter bed. The cracks occur mainly where the nut is fixed in the walls and in severe cases can form in the middle of the bed. Muballs form when accumulations of solids and grains are not broken down during backwash, and tend to be more prevalent when polymers are used. Light mudballs rise to the surface and are relatively easily dealt with, but heavier ones tend to sit within the bed and can cause dead zones. Typical solutions are to insert probes into the filter depth and flush-water jets. Alternatives include pumping the media through an injector and in severe cases digging the filter out. Filter boil occurs when the backwash is started rapidly causing the sand layer to separate from the gravel. The sand bed then breaks in one or more places and the flow is diverted through the weak points in the bed causing the sand to appear to boil. Similar problems can occur with horizontal migration of the gravel layer producing paths of lower resistance, which generates dead zones.

6.3.2   Direct filtration

Direct filtration is a process where the sedimentation or flotation unit ahead of the filter process is removed, and coagulated/flocculated water is fed directly onto the filters. In some cases, the flocculation tank is also removed and floc growth occurs directly in the filter; this process is known as in-line filtration. The process requires the generation of pin point flocs that are filterable rather than large dense flocs for sedimentation. Consequently, target floc sizes are <100 μm.

The process is gaining popularity because it is low cost in terms of both capital and operating costs. The capital costs are reduced as no sedimentation or flotation process is required upstream and the operating costs are reduced as a lower chemical dose is required, less power is used due to no primary removal process and less sludge is generated. Such advantages are balanced in that there is a reduced response time to any rapidly changing conditions and the process is not suitable for waters high in either organics or turbidity.

The selection of appropriate waters to be treated by direct filtration is usually based on the required coagulant demand exerted. Typical dose rates for direct filtration are between 2 and 14 mg l^{-1} of alum (as $Al_2(SO_4)_3 \cdot 14H_2O$), plus in some circumstances 0.2–2 mg l^{-1} of cationic polymer. The exact choice of dose and whether to use polymer is based on acceptable filter run times. As coagulant or polymer dose increases, the duration of the filter run will decrease as more load is being placed on the bed. Consequently, doses are minimised such that acceptable filter quality rather than best quality is often achieved. Target run times can be as low as 14–16 hr and this equates to 4% of the production being used for back-wash. For example, an alum concentration of 12 mg l^{-1} at a filtration rate of 12 m hr^{-1} will give a filter run time of about 16–20 hr. Typical filtration rates are reported between 2 and 15 m hr^{-1} although this is thought to be a little conservative as pilot studies regularly report that rates as high as 30 m hr^{-1} are appropriate.

The technology is most commonly used for relatively good quality waters which need additional colour or turbidity removal or have a dissolved metal ion, such as iron or manganese, to be removed. Typical water qualities treated by direct filtration are turbidities of up to 10 NTU, algae concentrations below 2000 asu ml^{-1} and organics concentrations of up to 6 mg l^{-1} (see Chapter 3).

6.3.3 Slow sand filtration

Slow sand filters are the original form of the technology and involve the use of very fine sand and low filtration rates relative to a conventional sand filter. The combination of the sand size and filtration rate means that solids accumulate on the surface layer forming a biologically active layer known as the *schmutzdecke* or *dirty skin*. When the head loss becomes excessive, the filter is cleaned by draining the water level below the sand and physically removing the schmutzdecke and up to 50 cm of sand. The layer is replaced with new sand and the plant reinstated into production. Typical cycle times range between 1 and 6 months but can be as long as 1–2 years depending on the water quality. The technology was first used for widespread municipal treatment of water in England and continues to be successfully used in treatment plants fed from the river Thames in London. Renewed interest has been shown elsewhere due to its ability to treat cysts (especially *Giardia lamblia*) whilst remaining a simple and easy technology to use. For this reason it is used on small remote works treating relatively high quality water and is becoming common in developing countries.

Recommended design of slow sand filters includes sand sizes of between 0.1 and 0.3 mm at depths of up to 1 m with a gravel support layer of 0.3–0.5 m. Typical filtration rates are 0.1–0.2 m hr^{-1} with available head losses of up to 2 m (Visscher, 1990). The filters are cleaned by removing the schmutzdecke and the top layer of sand by scraping. The process is conducted manually for small systems but has been mechanically developed for larger beds. Typical labour requirements are up to 100 hr per 1000 ft^2 when conducted manually and 60 hr for mechanically assisted. The filters are brought into service by filling them with water, usually upflow so as to reduce the likelihood of air getting trapped within the filter. The filter is then operated at a reduced rate for between 3 and 10 days while the schmutzdecke ripens. For this reason cleaning cycles should be avoided during the winter months in very cold climates as the bed has a propensity to freeze during the restart process.

The key element of the technology is the *schmutzdecke* which forms over the initial few cycles as the microflora community develops. The ecosystem contains bacteria, protoza such as ciliates and rhizopods, rotifers and aquatic worms. A hierarchy develops in relation to depth within the filter such that the smaller organisms inhibit the upper layers and are preyed upon by the larger organisms which range deeper into the bed. The schmutzdecke provides both physical and biological treatment and is the main source of removal in the bed. Additional treatment occurs through the top layer of the sand with the role of ciliates thought to be key to the process' bacterial reductions. Effective treatment does not occur until the schmutzdecke has formed and this takes from about 2 days up to 2 weeks depending on the water quality and local weather conditions. If cysts are a potential problem the water is often filtered to waste until the schmutzdecke becomes active.

Slow sand filters are not successful at treating river waters with high clay contents, as the colloidal clay penetrates deep into the bed and makes cleaning very difficult; or waters with high colours as the molecules are reasonably soluble and non-biodegradable and so do not get removed by any of the operating mechanisms. Surveys in the US have reported that typical average influent turbidities are usually less than 2 NTU and always less than 3 NTU to provide an appropriate level of treatment. Run times varied from 42 to 60 days in winter, which rose to 6 months during the summer period.

Table 6.5 SWTR assumed log removals (adapted from Cleasby & Logsdon, 1999).

Process	Log removals *Giardia*	Virus	Turbidity requirement (NTU)
Conventional	2.5	2.0	≤ 0.5 in 95% of samples, never >5
Direct	2.0	1.0	≤ 0.5 in 95% of samples, never >5
Slow sand	2.2	2.0	≤ 1 in 95% of samples, never >5

6.4 Applications

The two main areas of application for filtration processes are the reduction in turbidity and the protection against cysts, pathogenic bacteria and viruses. In practice, a well operated depth filter will routinely produce a final water turbidity of <0.1 NTU during phase 5 (*effective filtration*) of its operational cycle. Removal of other constituents relates more to the pre-treatment optimisation than the actual filtration process, as it will not remove dissolved organic or inorganic components unless they have been precipitated.

The removal of micro-organisms by filtration techniques is of great importance in the control of waterborne outbreaks caused by *Giardia lamblia* and *Cryptosporidium parvum*. The efficacy of the technology in this regard has resulted in the US Environment Protection Agency (USEPA) issuing a surface water treatment rule (SWTR) which species assumed log removals based on properly operating filters (Table 6.5). Successful removal of micro-organisms can be maximised by maintaining the effluent turbidity below 0.1 NTU, and assuming that 0.2 NTU represents breakthrough such that the filter should be backwashed. Filter ripening stages should also be minimised and preferably kept to below 1 hr (Cleasby & Logsdon, 1999).

References

Cleasby, J.L. & Logsdon, G.S. (1999) Granular bed and precoat filtration. In: R.D. Letterman, ed. *Water Quality and Treatment*. McGraw-Hill Inc, New York.

Droste, R.L. (1997) *Theory and Practice of Water and Wastewater Treatment*. John Wiley & Sons, New York.

Elimelech, M., Gregory, J., Jia, X. & Williams, R. (1995) *Particle deposition and aggregation— measurement, modelling and simulation*. Butterworth Heinmann, Oxford.

Visscher, J.T. (1990) Slow sand filtration: design, operation and maintenance. *J. AWWA.*, **82**(6), 67–71.

Membrane Processes 7

7.1 Introduction

Membrane processes represent an important subset of filtration processes as there are very few pollutants found in water that cannot be removed economically by membrane technology. A membrane, or more properly a semi-permeable membrane, is a thin layer of material containing holes, or pores, which allows the flow of water but retains the suspended, colloidal and dissolved species within the flow (depending on the size of the holes). The separation is based on the physical characteristics of the pollutants to be removed such as their size, diffusivity or affinity for specific contaminants. The membrane processes with the greatest application to potable water treatment are reverse osmosis (RO), nanofiltration (NF), ultrafiltration (UF) and microfiltration (MF).

The selection of microfiltration (0.1 μm), ultrafiltration (0.01 μm), nanofiltration (0.001 μm) or reverse osmosis (0.0001 μm) relates to a decreasing minimum size of the component rejected by the membrane (Figure 7.1). RO membranes are principally used to remove dissolved salts from brackish or salt water and nanofiltration membranes are generally used to remove disinfection by-product precursors such as natural organic matter (NOM). The larger pore size of microfiltration and ultrafiltration membranes means they are generally used to remove larger pollutants such as turbidity, pathogens and particles. Typically, the cost of membrane treatment increases as the size of the pollutant decreases, and as such the largest pore size membrane that can robustly achieve the separation is usually adopted. No commercially successful membrane exists to remove uncharged inorganic molecules such as hydrogen sulphide and small uncharged organics.

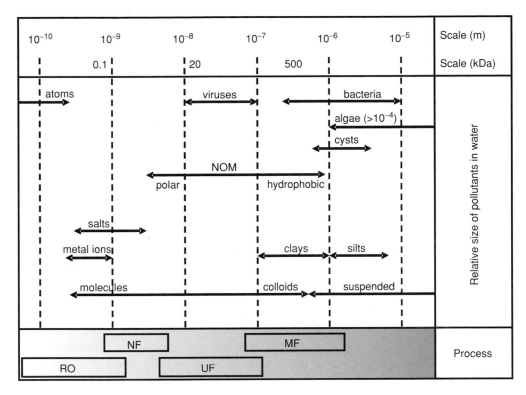

Fig. 7.1 Membrane separation overview.

In all four cases the bulk water passes through the membrane material under an applied pressure, leaving the pollutants in a concentrated form on the feed side of the membrane. Porous membranes (MF and UF) operate like sieves achieving separation mechanically by size exclusion. The retained material may either be dissolved or suspended depending on its relative size to that of pores within the membrane structure. Whereas dense membrane processes (NF, RO) operate in relation to the physicochemical properties of the permeating components and the membrane material.

Porous membranes are generally rated according to the size of the material they should retain. In the case of MF membranes this relates to a characteristic pore size, typically in microns (μm), whereas ultrafiltration membranes are more commonly rated in relation to a molecular weight cut-off in daltons (grams per mole). The actual physical size of the pores in dense membrane processes is less important as other mechanisms define the level of rejection that will occur. In these cases the membranes are rated according to their actual demonstrated rejection of a defined species: NF are generally rated in relation to multi-valent ions whereas RO membranes are rated in relation to univalent ions. Some membranes exhibit properties that can be associated with more than one type and as such the boundaries between the different membrane types is not sharp.

Membrane plants were first used in the UK around 12 years ago and were predominately small NF plants for colour removal in remote communities (Table 7.5). More recently, the uptake of membranes has risen sharply with more than 1200 MLD produced through the use of membranes today, compared to 10% of that figure for municipal wastewater treatment. A large proportion of the uptake is from a small number of large MF/UF plants for control against *Cryptosporidium*. This is a direct result of the UK Drinking Water Inspectorate (DWI) deciding that plants capable of continuously removing 0.1 μm particles do not need to regularly monitor *Cryptosporidium*, which is a difficult and expensive process. In the US a similar situation was developed through the Safe Drinking Water ACT (SDWA), which requires that surface water plants must achieve a 4 log (99.99%) removal/inactivation of enteric viruses. This is coupled with a tight standard of disinfection by product formation which limits the application of chlorine on some sites (see Chapter 9). Amendments to the rules, including more direct reference to *Cryptosporidium*, require treatment plants to meet a total credit of 2.5 log removal (of crypto); which can be met solely by a properly operating membrane plant.

This chapter will cover the process science and design of the membrane process as used in potable water production including a discussion of the key factors that influence operation and performance of the process. The chapter will then describe the main technology options including typical performance data for the main application areas.

7.2   Process science

7.2.1   *Process definitions*

In membrane processes there are three possible streams (Figure 7.2). A *feed* which passes through the membrane to produce a purified stream, the *permeate*. If the feed flow is perpendicular to the surface of the membrane the process is called dead-end filtration and the rejected material builds up on the surface of the membrane to form a cake. Alternatively, if the feed flow path is parallel to the membrane surface the process is called cross-flow filtration. In these cases the rejected material is concentrated in the flowing liquid, the *retentate*, which circulates around a loop reducing the build

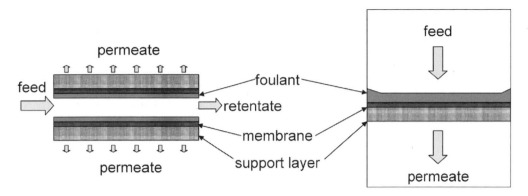

Fig. 7.2 Dead-end and cross flow configurations.

up of material on the membrane surface. Consequently, the amount of material retained in a dead-end filtration process is expected to be proportional to the amount of water passed through the membrane. In the case of cross-flow filtration, deposition continues until a balance is reached between material adhering to the membrane and the scouring effect of the cross-flow velocity re-entraining material into the flow (Figure 7.3). Whilst in principle this should generate a steady state, only a stabilised condition is possible due to the unavoidable deposition or adsorption of other fouling material.

The performance of any membrane system is defined in terms of the hydraulic throughput per unit area or flux and the degree of conversion. The flux, or permeate velocity as it is sometimes known, has the SI unit $m^3 \, m^{-2} \, s^{-1}$, or simply $m \, s^{-1}$. Other non-SI units are $l \, m^{-2} \, hr^{-1}$ (or LMI I) and $m^3 \, day^{-1}$ which tend to give more accessible numbers. The conversion or recovery, Θ, is the proportion of the feed that is passed through

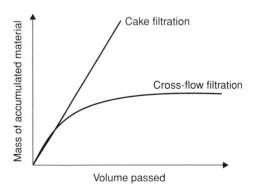

Fig. 7.3 Accumulation of material in dead-end and cross-flow filtration.

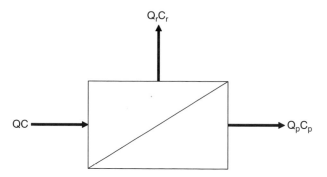

Fig. 7.4 Membrane mass balance.

the membrane as permeate and the rejection, R, is the amount of the pollutant material that is retained by the membrane. In relation to a mass balance across a membrane process these two properties can be defined as (Figure 7.4):

$$\Theta = \frac{Q_p}{Q}$$

$$R = 1 - \frac{C_p}{C}$$

The driving force for the process is the pressure differential or transmembrane pressure (TMP), between the feed and permeate sides of the membrane. The flux and transmembrane pressure are interrelated and as such either one may be fixed. Conventional practice is to fix the flux as the process normally needs to deliver a set quantity of water. The actual flux value used depends on the membrane system and the natural fouling propensity of the feed water.

7.2.2 Fouling and its control

During the operation of a membrane process the rejected material accumulates at the surface, or within the internal pores, of the membrane. This action is know as fouling and results in an increase in the hydraulic resistance and hence the required TMP (constant flux operation). Fouling can take place due to a range of physicochemical and biological mechanisms and should be distinguished from clogging which relates to the space in the membrane channels being blocked.

Once a membrane is fouled it must be cleaned by either backflushing or chemical cleaning to reduce the TMP of the system (Figure 7.5). Backflushing involves reversing the flow of permeate to dislodge any accumulated, and loosely bound, material. Backwashing usually occurs on a regular

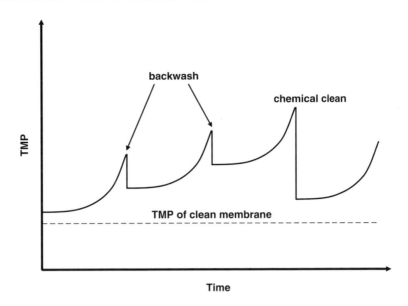

Fig. 7.5 Impact of backwashing and chemical cleaning.

cycle of the order of tens of seconds every 10–30 min and equates to a loss of between 5 and 10% of the production. Backflushing will not completely return the TMP to its original level and after a number of cycles the membrane requires a more extensive clean with chemicals to remove the tightly bound materials from the membrane's surface. Chemical cleaning involves, in general, an extended period during which the membrane is soaked in the cleaning solution. Most membrane plants are designed to enable cleaning in place (CIP) where the cleaning chemicals are circulated around the feed loop of the membrane plant via a separate set of pumps and tanks. In the case of dead-end systems the cleaning solution is passed back through the membrane by reverse flow. Chemical cleaning tends to occur only once the membrane has become severely fouled, although in recent applications a less severe but regular maintenance clean is being adopted to manage the build up of foulants and avoid lengthy soaking periods.

The main groups of foulants are (Table 7.1):

- Suspended and colloidal matter
- Scalants
- Micro-organisms
- Organic matter.

Solids
Solid foulants take the form of silts, colloidal clay, iron colloids and organic material. On the whole, gross solids are easily controlled by pre-filtration

Table 7.1 Common foulants and their control (adapted from Judd & Jefferson, 2003).

Foulant	Effect	Pre-treatment	Cleaning
Solids	Decreased permeability	<2 mg l^{-1}: cartridge filter >2 mg^{-1} depth filtration	Backwash
Colloids	Decreased permeability	Coagulation, depth filtration, MF/UF ahead of RO	Backwash
Scales	RO: decreased rejection	Threshold inhibitor (LSI < 2)	Hydrochloric acid (pH 4)
CaCO$_3$	Decreased permeability	Acid addition (LSI > 2) Ion exchange	2% citric acid ammoniated to pH 4
Metal oxides (RO)	Decreased rejection and permeability	Precipitation (aeration) and filtration	2% citric acid ammoniated 1% sodium hyposulphite
Active silica (RO)	Decreased rejection and permeability	Lime softening, alkali dosing and heat	Caustic soda (pH 11)
Dissolved organics	Decreased permeability	Coagulation, oxidation or adsorption	Caustic soda (pH 11) Alkaline detergents
Biological matter	Decreased permeability	Coagulation, oxidation or adsorption	1% formaldehyde hypochlorite

using cartridge or depth filters. Whereas colloidal fouling can be a severe problem and generally comes about due to agglomeration of the colloidal material as it passes through the membrane. The fouling propensity of water in relation to its solids content is most commonly determined by the fouling index or silt density index (SDI). Standard commercial test kits are available which involve passing the water through a standard 0.45 μm filter at a set pressure. The time taken to pass a given volume of water, normally 100 ml, is measured before (t_i) and after (t_f) a pre-determined operating period (t_t), say 15 min. The SDI is given by

$$SDI = \frac{100\left(1 - \dfrac{t_i}{t_f}\right)}{t_t}$$

RO and NF suppliers will typically specify a maximum SDI of around 5 for operation of their plant. A guide can also be determined by measuring the turbidity or turbidity to suspended solids ratio which provides a useful insight into the nature of the solids. Water above the target level must be pre-treated by either coagulation, MF/UF or a combination. Cleaning of solid foulants is most commonly achieved by backflushing as they do not tend to adhere to the membrane surface that strongly.

Scalants

Scalants are low solubility salts which precipitate onto the surface of the membrane as they increase in concentration due to their rejection. In potable water production, scale formation issues are mainly related to dense membrane processes and result in a reduction in both permselectivity and permeability. Scale formation is related to the solubility product of the given salt and represents the maximum value of the product of the molar concentrations of the two component ions of the salt. If the solubility is exceeded a precipitate is formed, although in practice the ionic product should not exceed 80% of the solubility limit to manage the problem effectively. The most common way to monitor the propensity for calcium carbonate scale formation is through the Langelier saturation index (LSI) which is defined as

$$LSI = pH - pH_s = pH - [(pK_2 - pK_s) + pCa + palk]$$

where pH is the measured value of the water and pH_s is the calculated form from a thermodynamic standpoint. Effectively, the LSI measures the degree to which the water is either super-saturated or under-saturated with calcium carbonate. Negative values indicate a corrosive water which will dissolve calcium carbonate off surfaces and a positive value indicates the potential for scale formation. The LSI does not work well between values of −1 and +1 when alternative indicators such as the Ryznar stability index (RSI) must be used.

The simplest way of preventing scale is to operate at a sufficiently low recovery that the concentration of the potential scalants never exceeds the solubility limit. However, there are practical limitations of such an approach which mean that other methods are commonly used. In the case of calcium carbonate scaling, or other hydrolysing precipitates, acid dosing can be used to adjust the LSI and avoid scale formation. Alternatively, scale inhibitors are used which interfere with the precipitation process and at the very least delay the onset of significant agglomeration until the water has left the membrane process. Chemicals include sodium hexametaphosphate (Calgon®), polyacrylates and polymalonates which block crystal growth sites by adsorbing onto the nanoscopic nuclei. Cleaning of fouled membranes usually involves acid cleaners such as low cost inorganic acids (sulphuric, hydrochloric, nitric) or organic acids (citric, acetic). The stronger the acid, the better its cleaning properties but it is also more corrosive so a balance is required, especially in terms of pH control.

Micro-organisms

Micro-organisms are ubiquitous, can survive in very low nutrient environments and even utilise pre-treatment chemicals such as flocculants, scale inhibitors (Ahmed & Alansari, 1989) and even by-products formed after the addition of chlorine (Applegate *et al.*, 1986). Consequently, biofilm formation is unavoidable and the main emphasis is towards control rather than prevention. To this end biocides are often used, the most common of which is chlorine in a sodium hypochlorite form adjusted to pH 10–11 to reduce its aggressiveness towards the membrane and the rest of the plant. Typical concentrations range between 10 and 100 mg l^{-1} although a 2–3% formaldehyde solution can also be used. The use of ceramic membranes offers advantages in this regard as they are capable of withstanding much more severe conditions and enable more effective cleaning to be achieved.

Organics

The key organic foulants are proteins, carbohydrates and natural organic matter. NOM is by far the most important organic foulant in potable water applications and represents a combination of a wide range of organic types varying in size, polarity and charge, making it very difficult to pre-treat the water sufficiently to eliminate any potential problems. Two strategies are generally used: coagulation of the charged material followed by MF/UF, or direct filtration with NF membranes. Organic foulants are most effectively cleaned by alkaline chemicals, primarily caustic soda. The solution generally needs buffering with bicarbonate or phosphate to maintain the pH and can contain surfactants to enhance their effectiveness.

7.3   Membrane integrity

A key issue with all membranes, and especially HF systems due to their uptake, is membrane integrity, which relates to the breaking of fibres which negates the barrier properties of the membrane. The loss of integrity is caused by either extraneous particles in the feed, air hammer from trapped air, loss of mechanical strength or inadequate fibre repair. Consequently, membrane integrity monitoring becomes an integral part of the process. There are two basic types of integrity test:

1. Particle challenge test
2. Pressure decay test (PDT).

Particle challenge tests are conducted online where small amounts of activated carbon or phage, sized to represent the target pollutant (i.e. normally AC in the size range 4–5 μm for *Cryptosporidium*) are inserted into the feed end and then the permeate is monitored for particle spikes at the given size band. However, it is far more common to use an offline PDT where the membrane is either pressurised with air (diffuse air flow) on the retentate or liquid side, or vacuumated on the permeate side, and the rate of pressure change monitored. This approach is favoured as there is a logarithmic correlation between the micro-organism rejection level, as defined by a log removal value (LRV), and the pressure decay rate (PDR) (kPa min^{-1}). Standard practice is to have an LRV target of four, beyond which the membrane module must be taken out of service for repair. The module is repaired by identifying the individual fibres that are broken through a submerged bubble test, and a pin is inserted at the exposed end of any broken fibre and the element refitted.

7.4 Process description

There is a wide variety of models in existence that attempt to predict either a steady state flux, transient flux decline (or TMP increase) or changes in the rejection character (Judd & Jefferson, 2003). However, there are essentially two approaches to the problem: cake filtration and film theory. In reality, the heterogeneity and diversity of components within the feed waters reduces the potential for the formation of generic equations describing filtration. The exception is in dense membrane operations with no complicating solid–liquid interactions where the effects of concentration polarisation can be predicted.

7.4.1 Cake filtration

The simplest description of membrane operation, and most commonly used for porous membranes (MF, UF), relates the flux of the system to the driving pressure by the use of a standard cake filtration expression:

$$J = \frac{\Delta P}{\mu R}$$

where J is the flux (m s^{-1}), ΔP is the transmembrane pressure (kg m^{-1} s^{-2} or Pa), μ is the fluid viscosity (kg m^{-1} s^{-1} or Pa s) and R is the total resistance to filtration (m^{-1}). The resistance term can be sub-divided into the individual components that generate it:

$$R = R_m + R_C = R_m + R_{rev} + R_{irrev}$$

where R_m is the hydraulic resistance of the clean membrane (Table 7.2), R_c is the resistance due to fouling, R_{rev} represents reversible fouling removed by backflushing and R_{irrev} represents irreversible fouling only removed by chemical cleaning. Consequently, this approach requires knowledge of

Table 7.2 Approximate values of membrane resistance and TMP for the different membrane processes.

Process	Approximate R_m (m^{-1})	Approximate TMP (bar)
RO	10^{10}	8–80
NF	10^8	3.5–10
UF	10^7	0.5–7
MF	10^6	0.3–3

the resistance of the membrane and the fouling layers, which is normally obtained empirically through a series of short-term experiments on new and fouled membranes with pure water before and after backwashing and chemical cleaning, respectively.

7.4.2   Film theory

The basis for film theory is the well-established concept of concentration polarisation (CP) which describes the tendency of solutes to accumulate near the surface of the membrane. The extent of CP directly relates to the size of the foulant species and as such is only of importance for dense membrane processes. The exact concentration profile of material towards the surface of the membrane is dependent on the balance of forces transporting material towards and away from the surface of the membrane. Assuming a one-dimensional system, the flux can be defined as

$$J = \frac{D}{\delta} \ln \left(\frac{C^*}{C} \right)$$

where D is the diffusion coefficient ($m^2\ s^{-1}$), C^* and C represent the concentrations at the membrane surface and in the bulk solution and δ represents the thickness of the boundary layer. The above expression does not contain a pressure term, although changes in the required pressure can be calculated via the resultant change in the osmotic pressure generated by the accumulation of solute material:

$$J = \frac{K_w}{\lambda} (\Delta P - \Delta \pi)$$

where K_w is the water permeability through the membrane and ΔP and $\Delta \pi$ represent the applied transmembrane and osmotic pressures, respectively. The osmotic pressure is related to the concentration of any given ion at the surface of the membrane by the van't Hoff equation:

$$\pi = y RT \sum c_i$$

where y is the osmotic coefficient to account for non-ideal behaviour and takes a value between 0.7 and 1. T is the absolute temperature and Σc_i is the sum of the solute ions.

7.5   Technology options

Membrane technology can be classified in a number of ways such as

1.   Approximate size of the removed species (see Section 7.1)
2.   The membrane material
3.   Configuration.

7.5.1 Materials

Membrane materials are divided into organic (polymer) or inorganic (ceramic or metallic) systems. Inorganic systems offer the advantages of lower inherent fouling propensities and perhaps more importantly, a much greater tolerance to chemical additions making them easier to clean. However, in water treatment applications the vast majority of systems are made from polymer materials due to the low cost of construction ($1 m^{-2}) as opposed to the higher costs of ceramic materials ($1000 m^{-2}).

To reduce the hydraulic resistance of most membrane materials the structure is anisotropic in that the pore size varies with depth through the material. A thin selective layer, normally 2–5 μm thick, sits on top of a very porous support material which provides the necessary mechanical strength (Figure 7.6). This concept is taken further in the case of some reverse osmosis membranes which have an additional ultrathin active layer, less than 0.5 μm, attached to the anisotropic support material to give improved selectivity.

The range of materials and methods of production used in the generation of membranes is quite diverse. Membranes can be produced by stretching, sub-atomic bombardment combined with etching and, in the case of ceramic materials, sintering (Judd & Jefferson, 2003). However, by far the most significant production method is by phase inversion. The production process involves dissolving the polymer in a suitable solvent, casting it into a film and then precipitating the polymer by adding another liquid, in which the polymer is sparingly soluble. The membrane skin forms at the interface between the two liquids and the pores are generated when the second liquid is removed. The desired characteristics of the membrane are generated by careful selection of the concentration, temperature and reaction time used.

Typical materials include cellulose acetate, polypropylene, polyetherimide and poly(vinylidene fluoride) (Table 7.3). The key issues are resistance to pH and organic solvents which are generally alleviated by using

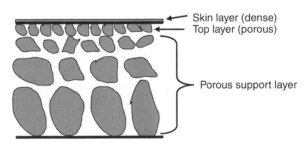

Skin layer (dense)
Top layer (porous)

Porous support layer

Fig. 7.6 Schematic and SEM of an anisotropic membrane.

Table 7.3 Phase inversion polymer membrane materials (adapted from Judd & Jefferson, 2003).

Polymer	Advantages	Disadvantages
Cellulose acetate (CA)	Chlorine resistant	Susceptible to pH>6
	Inexpensive	Limited thermal and chemical stability
	More fouling resistant than PA	Lower permeability
Polyamide (PA)	More stable than CA	Limited chlorine tolerance (<0.1 mg l^{-1})
	More permselective than CA	
Polyacrylonitrile (PAN)	Resistant to hydrolysis	Hydrophobic
	Resistant to oxidation	Needs copolymer to reduce brittleness
Polysulphone (PSU)	Very good all round stability	Hydrophobic
Polyether sulphone (PES)	Mechanically strong	
Poly(vinylidene fluoride) (PVDF)	Extremely high chemical stability	Highly hydrophobic
Polyetherimide (PEI)	High chemical stability	Hydrophobic
	Extremely high thermal stability	Less solvent resistant than PVDF
	Mechanically strong	Poorer alkaline stability than PSU or PAN
Polypropylene (PP)	Inexpensive	Hydrophobic

hydrophobic membranes. However, such surfaces have a high fouling propensity by non-specific adsorption of hydrophobic components in water, such as proteins and natural organic matter, greatly reducing hydraulic throughput. Consequently, much development work is involved in producing chemically and mechanically stable membranes with hydrophilic characters by modifying the standard materials. Modification processes include chemical oxidation to produce hydroxyl and carboxylic groups, chemical reaction, plasma treatment and grafting. A relevant example of this is the production of blends of sulphonated poly(ether ether ketone) (SPEEK) or poly (ether ether sulphone) (SPEES) with polysulphone or polyethersulphone for the production of highly negatively charged NF membranes for natural organic matter applications (Bowen *et al.*, 2001).

7.5.2   Configuration

The configuration, geometry and mounting arrangement, are important in determining the overall process performance. The optimum membrane configuration is one that has the following characteristics (Judd & Jefferson, 2003):

1. High membrane area to bulk volume ratio
2. Generates a high degree of turbulence as the water flows through it
3. A low energy expenditure per unit of permeate produced
4. A low cost per unit area
5. Easy to clean
6. Permits modularisation.

Table 7.4 Membrane configurations (adapted from Judd & Jefferson, 2003).

Configuration	Area/volume (m^2 m^{-3})	Cost	Turbulence promotion	Backflush possible?	Applications
Pleated cartridge	500–1500	Very low	Very poor	No	**MF**
Plate and frame	100–300	High	Fair	No	UF, RO
Spiral wound	800–1200	Low	Poor	No	UF, NF, **RO**
Tubular	150–300	Very high	Very good	No	UF, **NF**
Hollow fibre	10 000–20 000	Very low	Very poor	Yes	**MF**, RO

Note: main application marked in bold.

With the exception of modularisation, which all configurations permit, the other factors are generally mutually exclusive and as such a number of different configurations have been developed in order to provide a balance of the factors (Table 7.4). The pleated cartridge, plate and frame and spiral wound (SW) configurations are all based on planar geometries. Pleated cartridges are a low-cost design used exclusively in MF applications and which act as a disposable unit for polishing relatively pure water. Plate and frame systems are constructed in a similar way to heat exchangers and comprise a series of rectangular or circular plates stacked together between separators/support plates. The modules can be open or pressurised and are mainly used for UF and RO applications. Spiral wound systems, used mainly in NF and RO, comprise two membranes sandwiched together to form a sealed rectangular bag, on three sides, which is rolled up to form a cylinder. Spacers are used to maintain the desired gap between layers to allow the flow to feed through the module. An interesting recent development involves the production of ceramic membranes by extrusion, which greatly reduces costs by up to a factor of 10.

Tubular modules are multi-tube arrays with mechanical support for strength (Figure 7.7). They represent the lowest area to volume ratio but

Hollow fibre

Fig. 7.7 Example of a hollow fibre membrane used in potable water treatment.

provide the greatest turbulence generation. When the tube diameter is sufficiently large, i.e. >5 mm, it is possible to mechanically clean in place as with the FYNE® process developed by PCI membranes which uses foam balls to wipe the surface of the membrane. Hollow fibres are small diameter tubes which are self supporting enabling very simple module designs and very high area to volume ratios. The flow can be either out to in or in to out with the hydrodynamic control being especially poor in the former. Consequently, many HF systems backflush to control the build up of material on the membrane surface. A recent development is the use of submerged hollow fibre systems with aeration to aid turbulence promotion in the tank. Such systems are a direct transfer from wastewater treatment and the development of membrane bioreactors, as they provide a relatively inexpensive method of delivering large-scale membrane processes (Stephenson *et al.*, 2000).

In the case of dense membrane processes the conversion is limited by the membrane area and the rate of extraction to values no higher than 20%. Consequently, most systems use sequential stages to produce a higher conversion than is possible in a single stream (retentate staging) (Figure 7.8). For high salinity feed waters it is not desirable to use the retentate in latter stages as the osmotic pressure requirement will be excessive. Instead, the permeate is staged to improve the overall rejection of the plant (permeate staging).

7.6 Applications

A large number of membrane companies exist; however, in terms of water processing there are a relatively small number of core companies that

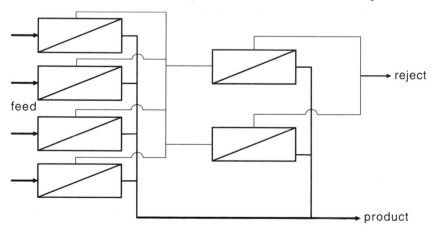

Fig. 7.8 Schematic of staging.

Table 7.5 Summary of the main commercial systems used in water treatment in the (MF-RO) UK and (MF-UF) USA (as of 2004).

Company	Systems/pore size	Materials	Configurations	Approximate total capacity (1000 m^3d^{-1})
Aquasource	UF (0.01 μm)	CA	HF, side stream	18
Kalsep/Hydr.	UF (150 kDa)	PES	HF, side stream	68
Koch/FS	UF (0.01 μm)	PS	HF, side stream	250
Koch/FS	NF (200–2000 kDa)	CA, PA	SW, side stream	11
Memcor/USF	MF (0.1–0.2 μm)	PP, PVDF	HF, submerged, side stream	1400
Norit/X-flow	UF (100–200 kDa)	PES	HF, side stream	400
Pall	MF (0.1 μm)	PVDF	HF, side stream	400
PCI	NF (400 kDa)	CT	T, side stream	12
Zenon	UF (0.02–0.04 μm)	PVDF	HF, submerged	450

Hydr. = Hydronautics.

supply the vast majority of membrane plants (Table 7.5). The majority of the plants, by total capacity, are side stream HF used for pathogenic organism control (*Cryptosporidium* etc.). There are also a large number of sites of small flow that are used for NOM control, which are mainly tubular NF and SW membranes (Table 7.6). Uptake of the systems has grown continually since about 1999, which is enabling the price per unit area to decrease and cost competitiveness of larger membrane sites to improve.

A detailed case site list is given for the UK as an example (Table 7.6). In 2004, there were 107 reported membrane plants for potable water production split between MF/UF (47) and NF/RO (36). The 107 plants produce $1\,236\,000$ m^3 d^{-1} of which $1\,197\,000$ m^3 d^{-1} is produced from MF/UF plants and 39 000 m^3 d^{-1} from NF/RO plants showing the difference in the plant sizes. The majority of the MF/UF plants have been installed since 2000 (for cyst control) and range from individual capacities of 8.5–160 000 m^3 d^{-1}. The majority of the plants are pumped-flow side-stream membranes with either PP or PES membranes operated roughly equally as in to out and out to in flow. A small number of submerged PVDF membranes have recently been installed although their application is much more established in the US.

Reported flux values from the UF plants vary between 40 and 120 LMH with an average flux of 60 LMH working at TMPs of 0.2–0.8 bar. No specific trend between feed water character and operating flux rate can be observed across all the sites except that in general, as expected, the permeability of the membrane appears to decrease as the feed water becomes more polluted. The majority of fouling incidents occur following a period of rainfall after a dry spell, mainly just after the summer has finished, and

Table 7.6 Site list of operational membrane plants of water treatment works in the UK (adapted from Judd, 2004).

Supplier	Site	Geometry	Type	Flow (m^3 d^{-1})	Date
Acumem/USF	Huntington	HF	UF	80 000	1998
Aquasource	Tosson	HF	UF	8000	2002
Aquasource	Tyn-y-Waun	HF	UF	10 000	–
Kalsep/Hydr.	Alderley	HF	UF	5000	2001
Kalsep/Hydr.	Allenheads	HF	UF	130	2000
Kalsep/Hydr.	Banwell	HF	UF	6000	2001
Kalsep/Hydr.	Carradale	HF	UF	360	–
Kalsep/Hydr.	Carrshield	HF	UF	80	2000
Kalsep/Hydr.	Charterhouse	HF	UF	2000	2001
Kalsep/Hydr.	Crumpwood	HF	UF	7400	2002
Kalsep/Hydr.	Forum	HF	UF	2000	2001
Kalsep/Hydr.	Frome	HF	UF	5000	2001
Kalsep/Hydr.	Halton-lea-gate	HF	UF	210	2000
Kalsep/Hydr.	Oldford	HF	UF	15 000	2002
Kalsep/Hydr.	Sherborne	HF	UF	4000	2001
Kalsep/Hydr.	Slaggyford	HF	UF	250	2000
Koch/FS	Backies	SW	RO	2200	1999
Koch/FS	Badachro	SW	RO	200	2000
Koch/FS	Bonar Bridge	SW	RO	1100	2000
Koch/FS	Bracadale	SW	RO	190	1996
Koch/FS	Broadford	SW	RO	705	1999
Koch/FS	Bunessan	SW	RO	550	1996
Koch/FS	Bunwell	SW	RO	400	1997
Koch/FS	Dagenham	SW	RO	4700	1997
Koch/FS	Dalmally	SW	NF	470	2000
Koch/FS	Debden Rd	SW	RO	6000	2000
Koch/FS	Flag Fen	SW	RO	1440	2000
Koch/FS	Inverasdale	SW	NF	200	2000
Koch/FS	Kilham	SW	RO	1500	1998
Koch/FS	Kyle	SW	NF	1700	2001
Koch/FS	Lochgair	SW	NF	70	1992
Koch/FS	Mallaig	SW	NF	1500	2002
Koch/FS	Nth. Pickenham	SW	RO	1600	1997
Koch/FS	Rushall	SW	RO	2400	1997
Koch/FS	St Aidans	SW	RO	2000	1996
Koch/FS	Teangue	SW	NF	350	2001
Memcor/USF	Batchworth	HF	SMF	41 000	2003
Memcor/USF	Chertsey	HF	UF	40 000	2003
Memcor/USF	Corfe Mullen	HF	MF	28 000	2001
Memcor/USF	Cornhow	HF	MF	35 000	2000
Memcor/USF	Crathie	HF	MF	240	–
Memcor/USF	Dorney I	HF	MF	27 000	2003
Memcor/USF	Dorney II	HF	MF	29 000	2004
Memcor/USF	Drellingore	HF	MF	14 000	2003
Memcor/USF	Ennerdale	HF	MF	59 000	1999
Memcor/USF	Farlington	HF	MF	84 000	2003
Memcor/USF	Friar Waddon	HF	MF	10 000	2001
Memcor/USF	Homesford	HF	MF	65 000	1999
Memcor/USF	Invercannie	HF	SMF	72 000	2004
Memcor/USF	Lovedean	HF	UF	12 100	2004
Memcor/USF	Lye Oak	HF	MF	7500	2002
Memcor/USF	Otterbourne	HF	SMF	58 000	2004
Memcor/USF	Ottinge	HF	MF	3500	2002
Memcor/USF	Washpool	HF	MF	11 000	2003
Memcor/USF	Whitebull	HF	UF	45 000	1998

Table 7.6 (*Continued*)

Supplier	Site	Geometry	Type	Flow (m³ d⁻¹)	Date
Memcor/USF	Winwick	HF	UF	16 000	1997
Norit/X-flow	Brynberian	HF	UF	1000	2002
Norit/X-flow	Chalford	HF	UF	13 800	2002
Norit/X-flow	Clay Lane	HF	UF	160 000	2001
Norit/X-flow	Coombe	HF	UF	1000	2004
Norit/X-flow	Crug	HF	UF	3000	2001
Norit/X-flow	Dolbenmaen	HF	UF	5000	2002
Norit/X-flow	Gwastadgoed	HF	UF	1000	2002
Norit/X-flow	Hatton	HF	UF	2003	2002
Norit/X-flow	Hook	HF	UF	2400	2002
Norit/X-flow	Houndall	HF	UF	4500	2001
Norit/X-flow	Inverness	HF	UF	34 000	2002
Norit/X-flow	Keldgate	HF	UF	90 000	2001
Norit/X-flow	Kepwick	HF	UF	2000	1998
Norit/X-flow	Llandium	HF	UF	24 000	2002
Norit/X-flow	Lydbrook	HF	UF	4000	2002
Norit/X-flow	Nth. Mymms	HF	UF	36 000	2002
Norit/X-flow	Nth. Orpington	HF	UF	9100	2001
Norit/X-flow	Pen-Y-Bont	HF	UF	2500	2002
Norit/X-flow	Trecastell	HF	UF	3500	2002
Norit/X-flow	West Stonedale	HF	UF	800	2002
Norit/X-flow	Wilmington	HF	UF	3000	2001
PCI	Achnasheen	T	NF	50	2001
PCI	Ardeonaig	T	NF	100	2002
PCI	Ardfern	T	NF	110	2000
PCI	Arinagour	T	NF	42	1999
PCI	Ballygrant	T	NF	150	1998
PCI	Balquhidder	T	NF	100	1998
PCI	Brif O'Turk	T	NF	150	1999
PCI	Carrick Castle	T	NF	70	1998
PCI	Carsphairn	T	NF	47	2001
PCI	Cladich	T	NF	20	1994
PCI	Colonsay	T	NF	123	2000
PCI	Corric	T	NF	190	1999
PCI	Craighouse	T	NF	80	2001
PCI	Crianlarich	T	NF	250	1999
PCI	Dalwhinnie	T	NF	150	2001
PCI	Dervaig	T	NF	150	1997
PCI	Eredine	T	NF	25	1996
PCI	Gigha	T	NF	140	2002
PCI	Gorthleck	T	NF	443	1998
PCI	Ilseornsay	T	NF	150	2001
PCI	Kilberry	T	NF	10	1998
PCI	Killin	T	NF	200	1999
PCI	Kirkmichael	T	NF	120	1995
PCI	Laggan Bridge	T	NF	20	2001
PCI	Lochearnhead	T	NF	300	2002
PCI	Lochgoilhead	T	NF	136	2001
PCI	Lochranza	T	NF	310	1999
PCI	Raasay	T	NF	200	2001
PCI	Strollamus	T	NF	100	2001
PCI	Tomnavoulin	T	NF	100	1995
Toray	Jersey	SW	RO	6000	1997
Toray	Scilly Ilses	SW	RO	100	–

HF = Hollow fibre; SHF = Submerged hollow fibre; SW = Spiral wound; Tube = Tubular.

are commonly associated with turbidity spikes. Backflush is on average conducted every 60 min for between 30 and 120 s. Chemical cleaning tends to be staged with 1 acid clean with sulphuric and hydrochloric acid for every 3–4 caustic cleans and hypochlorite every 3–14 days. In the case of the NF plants reported flux rates range between 10–30 LMH at TMPs of up to 6 bar operating at temperatures between 3–20°C.

References

Ahmed, S.P. & Alansari, M.S. (1989) Biological fouling and control at RAS Abu Jarjur RO plant – a new approach. *Desalin,* **74,** 69–84.

Applegate, L.E., Erkenbrecher, C.W. & Winters, H. (1986) Monitoring control of biological activity in Permasep seawater RO plants. *Desalin.,* **65,** 331–359.

Bowen, W.R., Doneva, A.D. & Yin, H.-B. (2001) Separation of humic acid from a model surface water with PSU/SPEEK blend UF/NF membranes. *J. Membr. Sci.,* **206,** 417–429.

Judd, S.J. & Jefferson, B. (2003) *Membranes for Industrial Wastewater Recovery and Reuse.* Elsevier Science, London.

Judd, S.J. (2004) Membrane database. http://www.main.wizzy.co.uk/ (11th April 2005).

Stephenson, T., Brindle, K., Judd, S.J. & Jefferson, B. (2000) *Membrane Bioreactors for Wastewater Treatment.* IWA Publishing, London.

Adsorption Processes 8

8.1 Introduction

Adsorption was first used (for medicinal purposes) as early as 1500 BC in Egypt when charcoal was used to adsorb odorous vapours from putrefying wounds and from within the intestinal tract. In the eighteenth century carbons derived from blood, wood and animals were used for the purification of liquids. All of these materials however were only available as powders and had to be mixed with the liquid to be treated which, following the prescribed contact time, then required a method of liquid separation. Percolation technology, when the liquid to be treated is continuously passed through a column using a granular form of bone char, was first used in England by the sugar industry. Processes were not developed until the first part of the twentieth century to produce activated carbons with defined properties. The manufacturing process using either steam or chemical activation could however still only produce a powdered form of carbon. In the 1930s a manufacturer successfully developed a coal based granular activated carbon with a substantial pore structure and good mechanical hardness. The ability to reactivate this type of granular carbon has drastically reduced the cost of adsorption as a unit process. Table 8.1 compares the characteristics of commonly used adsorbents in the treatment of water.

This chapter will focus on activated carbon adsorbents used for potable water production and covers the process science and design and its applications. The application of adsorbents for inorganic removal will be discussed further in Chapter 11.

8.2 Process science

The process of adsorption involves separation of a substance, termed an adsorbate, from the liquid phase, and the concentration at the surface of

Table 8.1 Comparison of commonly used adsorbents in water treatment.

Adsorbent	Mean particle size (mm)	Surface area $(m^2\ g^{-1})$	Density $(g\ cm^{-3})$	pzc	Applications
GAC	1	500–2000	0.5	4	Taste and odour, pesticides, NOM removal
PAC	0.1	500–2000	0.5	4	Taste and odour, pesticides
Activated alumina	11	220	0.91	7.3	Arsenic and fluoride removal
β-FeOOH	0.03	142	0.45	8.3	Arsenic removal

a material termed an adsorbent. The process follows four phases. Initially the adsorbate must first travel from the bulk liquid phase to the liquid film surrounding the absorbent, typically a carbon particle. Second, the adsorbate must travel through the liquid film surrounding the carbon to the interstitial voids. Third, the adsorbate must diffuse through the carbon voids in the carbon solid phase and fourth, finally adsorb onto the carbon. The extent of adsorption depends on the specific nature of the carbon and of the molecules being adsorbed and is a function of the concentration, temperature and pH. Adsorption can be physical or chemical. In physical adsorption the impurities are held on the surface of the carbon by weak van der Waals forces and there is no significant redistribution of electron density in either the molecule or at the substrate surface. In chemisorption, a chemical bond involving substantial rearrangement of electron density is formed between the adsorbate and substrate. The nature of this bond may lie anywhere between the extremes of virtually complete ionic or complete covalent character. If the reaction is reversible, molecules accumulate on the surface until the forward reaction (adsorption) equals the rate of the reverse reaction (desorption).

A number of models have been derived in an attempt to explain and quantify the phenomenen of adsorption many of which stem from the work of Langmuir (Langmuir, 1918). Langmuir described the equilibrium between surface and solution as a reversible chemical reaction between the two phases, and that this reaction has a fixed free energy of adsorption, ΔG_a^0:

$$K_a = \frac{\Gamma}{[\Gamma_T - \Gamma]^C}$$

where

Γ_T = the total number of surface adsorption sites in mol per unit area
Γ = the number of sites occupied

$$K_a = e^{-\Delta G_a^0 / RT}$$
R = gas constant $(8.314 \text{ J}/(\text{mol} \cdot \text{K}))$
T = temperature [K].

In practice many of these quantities are unknown and Langmuir developed the above equilibrium into the isotherm:

$$\frac{q}{Q} = \frac{bC}{1 + bC}$$

where

q = the concentration of the adsorbate in mol g^{-1} adsorbent
Q = the adsorbent's saturation capacity for the adsorbate, mol g^{-1}
b = an empirical constant.

Rearranging Langmuir's equation gives:

$$\frac{C}{a} = \frac{1}{bQ} + \frac{C}{Q}$$

Plotting C/q against C gives a graph with a straight line with gradient $1/bQ$ and intercept $1/Q$. The Langmuir isotherm assumes that saturation is reached when the adsorbent is covered with a monolayer of adsorbate. Brunauer Emmett and Teller further developed the model to allow for multi-layers with each additional layer of adsorbate in equilibrium with the previous layer (Brunauer *et al.*, 1938). The principles are similar except that Ka is redefined as $K_a = e^{-\Delta G_v^0 / RT}$ where ΔG_v^0 is the free energy of precipitation of the adsorbate. The Brunauer Emmett and Teller (BET) isotherm is

$$\frac{a}{Q} = \frac{BC}{(C_s - C)\lceil 1 - (B - 1)(C/C_s) \rceil}$$

where

B = a dimensionless empirical constant
C_s = the saturation concentration of adsorbate in solution.

The Langmuir isotherm can, theoretically, be extended to a multi-component system as follows:

$$\frac{q_i}{Q} = \frac{b_i C_i}{1 + \sum_{i=1}^{n} b_i C_i}$$

This allows the equilibrium capacity of an adsorbent for a mixture of components to be calculated from single adsorbate data. However, the calculation should always be backed up by experimental verification. Both the Langmuir and BET isotherms assume a homogeneous surface with the energy of adsorption constant for all sites. In reality this is not the case, because surfaces are heterogeneous. Freundlich's work attempted to make allowance for this and he concluded that the isotherm should have the form

$$q = KC^{\frac{1}{n}}$$

where

K = an empirical constant
n = another empirical constant.

Thus, a log–log plot of surface concentration, q, against equilibrium concentration, C will be linear. The Freundlich isotherm is more widely applicable in water treatment applications than either the Langmuir or the BET isotherms.

8.3   Activated carbon

Activated carbon is manufactured from a wide variety of carbonaceous material such as coal, bituminous coal (lignite), bone, wood and coconut shell. The basic carbonaceous material is subjected to a manufacturing process which creates a huge internal area within the carbon particles. For example one gram of ground coal has an internal surface area of only 10 m^2; during the process of activation this internal surface area is increased up to 1000 m^2. Hence, the process of activation transforms the carbon particle into a useful adsorbent. The most common process of activation consists of thermal treatment which is carried out in two stages. First, the material is carbonised at about 700°C in the absence of oxygen to reduce the volatile content and convert the raw material to a char. This is followed by further heating up to 1000°C in a regulated oxygen and steam environment. The steam penetrates the carbon particle and selectively oxidises the internal carbon atoms to create the large internal surface area.

Within the furnace the main gaseous reactions can be represented as follows:

$$H_2O(g) + C(s) \Rightarrow CO(g) + H_2(g) \text{ endothermic}$$
$$CO_2(g) + C(s) \Rightarrow 2CO(g) \text{ endothermic}$$

then

$$CO(g) + H_2O(g) \Rightarrow CO_2(g) + H_2(g)$$

When using oxygen the reaction is exothermic and it is more difficult to control the attack on the carbon structure.

$$3/2\,O_2(g) + C(s) \Rightarrow CO_2(g) + CO(g)$$

There are two principle forms of activated carbon used in water treatment: granular activated carbon (GAC) and powdered activated carbon (PAC). Although both types can be produced from the same raw material their typical average particle diameters vary from greater than 1 mm for the GAC to less than 0.1 mm for PAC. This variation in size can affect the rate of adsorption but not the amount adsorbed, since the greater proportion of the particle's surface area lies within its internal structure (pores) and not on its external surface.

Carbons are generally categorised by pore size according to the classification below:

macropore >50 nm

mesopore 2–50 nm

micropore 1–2 nm

minimicropore <1 nm

Of the carbon sources it is known that coconut shell-based GAC has the highest surface area, generally over 1000 m^2 g^{-1}. Bituminous-based GAC has a relatively large surface area, approximately 900 m^2 g^{-1}, whilst lignite-based GAC has a surface area of approximately 650 m^2 g^{-1}. Lignite GAC is more suited to larger molecules whilst coconut shell-based GAC generally has a larger surface area than coal-based GAC, and a very large percentage of micropores. A comparison of carbon sources and activated carbon structures is shown in Figure 8.1.

The following main characteristics should be considered in respect of GAC specifications.

Total surface area is normally determined by measuring the amount of nitrogen adsorbed to produce an adsorption isotherm. This data is then analysed using the Brunauer Emmett Teller isotherm equation which assumes monolayer adsorption of nitrogen molecules only. The resulting surface area is given as the BET surface area, normally in units of m^2 g^{-1} of media.

Apparent density is normally expressed as the 'apparent bulk density'. In this instance only the density of the carbon in air is being measured. Typical values range from 350 to 500 kg m^{-3}.

True bulk density includes contributions from the carbon, the weight of water within the pore structure and the water attached to the carbon. Typical values range from 900 to 1300 kg m^{-3}.

ACTIVATED CARBONS COMPARATIVE CHART

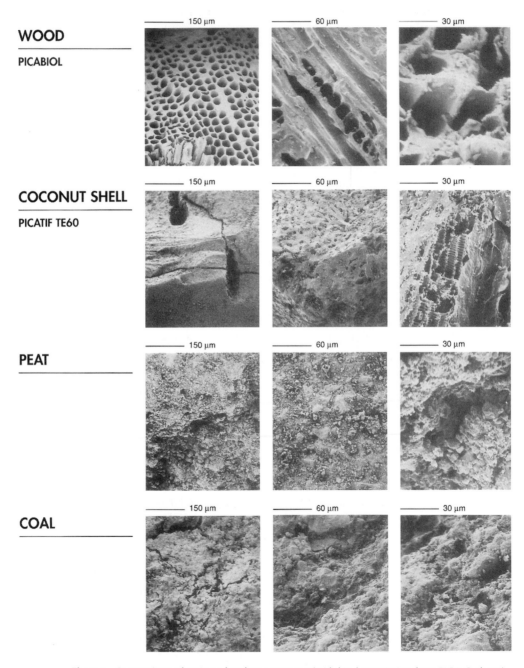

Fig. 8.1 Comparison of activated carbon structures (with kind permission from PICA Carbons).

Hardness is measured by placing a screened and weighed sample in a 'hardness pan' with a number of steel balls. This is then subjected to a combined rotating and tapping action for 30 min. The degree of particle size degradation is measured as the 'ball–pan hardness' and is calculated as follows:

$$H = \frac{100\,A}{B}$$

where

H = ball–pan hardness number
A = weight of sample retained on hardness sieve
B = weight of sample used in test.

Adsorption capacity is usually estimated by the adsorption of iodine or methylene blue. The iodine number is defined as the amount of iodine (mg) adsorbed by 1.0 g of carbon at a residual concentration of 0.02 N. The test results give an indication of the surface area available within the micropores. Methylene blue adsorption is predominantly within the mesopores and the methylene blue number is defined as the amount of methylene blue (mg) adsorbed by 1.0 g of carbon at a residual concentration of 1.0 mg l^{-1}.

When the effluent quality from a carbon contactor reaches minimum water quality standards, the spent carbon is removed for regeneration. Small treatment works usually find regeneration of their spent carbon at an off-site commercial reactivation facility to be the most convenient and economical method, whilst treatment works that contain at least 500 tonnes of carbon find on-site regeneration to be cost effective. Carbon regeneration is accomplished primarily by thermal means. Organic matter within the pores of the carbon is oxidised and thus removed from the carbon surface. The two most widely used regeneration methods are rotary kiln and multiple hearth furnaces. Approximately 5–10% of the carbon is destroyed in the regeneration process or lost during transport and must be replaced with virgin carbon. The capacity of the regenerated carbon is slightly less than that of virgin carbon. Repeated regeneration degrades the carbon particles until equilibrium is eventually reached providing predictable long-term system performance. The main elements of the regeneration process are:

1. Pre acid washing
2. Reactivation
3. Quenching and post treatment
4. Flue gas treatment.

An optional pre acid washing process is available which has the ability to wash the spent GAC with hydrochloric acid (pH 2) for several hours or until set process criteria are reached. The benefits of this are the removal of damaging inorganic compounds such as calcium which can catalyse the thermal degradation of the GAC structure and lead to reduced physical strength and shorter operational life. Additional benefits are also that the conditioning of the regenerated GAC on return to service is much faster due to the removal of inorganic species which can cause difficulties (Mn, Al etc). From the spent hopper the GAC is transported to the top of the furnace (using an eductor system) via a dewatering screw which removes up to 85% of the water. A typical rotary hearth furnace consists of six hearths (refractory lined) and is capable of processing up to 40 m^3 day^{-1}. A revolving central shaft with attached rabble arms, sweeps the carbon from the inlet port on the middle of the top hearth to the outside, where it drops onto the hearth beneath. It is then rabbled to the inside and falls to the next hearth and so on. The constant turnover action of the rabbles produces an intimate contact between the carbon granules and the furnace gases.

Factors that affect the quality of the regenerated product are

1. Oxidising gas
2. Furnace temperature
3. Residence time in the furnace
4. Concentration and type of substance adsorbed
5. Type and concentration of inorganic impurities adsorbed
6. Type of activated carbon (i.e. wood, coal based etc).

Regeneration of the carbon can have the following deleterious effects on the integrity of the structure:

• Small pores (<2 nm) may be lost, whilst larger pores are created
• The particle diameter may be reduced
• Hardness of the particle decreases.

The regeneration plant is governed by the Environmental Protection Act 1990 in the UK which requires that emissions from the plant meet certain standards. As a result it has been necessary for the installation of a number of process units to reduce the contaminants in the flue gas emissions from the furnace stack. Flue gas emission treatment firstly consists of an after-burner which provides for a retention time of 2 s at 8500°C in a minimum 6% oxygen environment, which is necessary for the oxidation of harmful organic products (e.g. dioxins). This is followed by passage through a heat exchanger to reduce the temperature of the gas. Particulates are then removed by passing through a venturi scrubber. A tray scrubber is then used to remove acidic gases. Finally, the temperature is increased to 1000°C and discharged to atmosphere via a chimney. It is important to increase the temperature to avoid a visible plume. Continuous monitoring and independent emission testing ensure compliance with the regulations. The excess heat removed by the heat exchanger can in part be reused.

8.4 Applications

Activated carbon is widely used in the water industry to remove organic and inorganic compounds. In particular it is used to remove (i) natural organic matter and colour, (ii) pesticides, (iii) taste and odour (iv) algal toxins. There are some general rules that can be used to predict if GAC is likely to remove a molecule:

- Larger molecules adsorb better than smaller molecules.
- Non-polar molecules adsorb better than polar molecules.
- Slightly soluble molecules adsorb better than highly soluble molecules.
- pH may have an influence on the extent of adsorption as it can control both polarity and solubility of a molecule.

Figure 8.2 compares the Freundlich isotherm constant, K for a range of problem water contaminants. The constant K is related to the capacity of the activated carbon for a specific molecule where the larger the value of K the larger the capacity will be and the molecule will be more easily adsorbed. From the data shown in Figure 8.2 it can be seen that activated carbon has a significantly larger capacity for the pesticide atrazine than it does for the taste and odour causing molecules, geosmin and 2-methylisoborneol (MIB). More information on the use of activated carbon to adsorb organic molecules is given in Chapter 10.

8.5 Adsorbers

Granular carbon is produced in the form of granules, typically around 1–2 mm diameter, which are used in deep bed contactors or 'columns'. As water passes down the column the carbon becomes exhausted from the top downwards. If we consider the 'exhaustion front', that is the interface between the carbon above and the fresh carbon below we see a concentration profile like that shown in Figure 8.3. It is an over-simplification

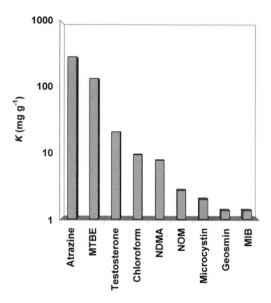

Fig. 8.2 Comparison of the Freundlich isotherm constant (K) for a range of contaminants (Awwa, 2000; Hall & Sheppard, 2005).

of the real situation, but we can assume that above the dotted line the carbon is 100% exhausted whilst below it the carbon is 100% fresh. At the interface there is a band of carbon through which the state changes and, at the bottom as the exhaustion front reaches the bottom of the bed, the sorbate will finally break through the bed.

The volume of carbon required in a treatment plant is usually determined by the empty bed contact time (EBCT) and is calculated as follows:

$$V = Qt$$

where

V = GAC volume (m^3)
Q = Flow rate $(m^3\,h^{-1})$
t = empty bed contact time in hours.

However, it is wise to check that the capacity is sufficient to give a reasonable run time, especially if the carbon is to be taken off site for regeneration:

$$T = \frac{1000\,V\rho\,C}{Qc}$$

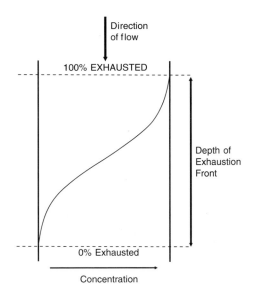

Fig. 8.3 The exhaustion front.

where

ρ = Carbon density (typically 0.5 te m^{-3})

c = Influent concentration of adsorbate mg l^{-1}

C = Carbon adsorption capacity for adsorbate mg g^{-1}.

In waterworks applications the carbon volume is usually large and consideration must be given to carbon handling. Most road tankers will take a carbon load of 20 m^3 so most GAC contactors are designed for a carbon volume of some multiple of 20 m^3. Examples of a pressure adsorber are shown in Figures 8.4 and 8.5.

8.6 Ozone/GAC

The combination of ozonation followed by granular activated carbon has become almost an industry standard for pesticide removal. Pesticides can be adsorbed directly on to activated carbon and there are several works that use this process alone. Where the problem is a seasonal one in surface water then it may be economic to use powdered activated carbon as an adsorbent but, more usually, granular activated carbon is used. Ozone is a strong oxidising agent by virtue of the generation in water of free hydroxyl radicals. These break down the large pesticide molecules into smaller ones which are readily adsorbed by activated carbon (GAC or PAC) or can be

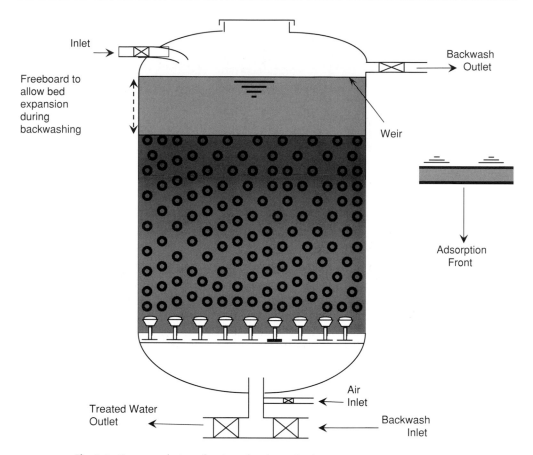

Inlet

Backwash
Outlet

Freeboard to
allow bed
expansion
during
backwashing

Weir

Adsorption
Front

Air
Inlet

Treated Water
Outlet

Backwash
Inlet

Fig. 8.4 Common design of activated carbon adsorber.

metabolised by bacteria growing on the surface of carbon granules ('biological activated carbon, BAC'). Ozonation upstream of carbon extends
the life of the activated carbon and, since both processes have high capital and operating costs, careful economic analysis is required. Although
ozonation is often installed specifically for pesticide removal, it also provides oxidation of organics (taste and odour removal), removal of THM
precursors and disinfection. Application of ozonation prior to coagulation (pre-ozonation) is often beneficial in reducing the coagulant demand
and oxidising inorganic species and organic algae, whilst post-coagulation
(downstream of filtration) is more effective at pesticide removal and disinfection. Ozone residuals are unstable so residual chlorination is normally
applied prior to distribution. Ozone has been used in water treatment
for disinfection and taste and odour control since the early part of the
twentieth century.

Fig. 8.5 Example of activated carbon pressure adsorbers (image courtesy of Grafham Carbons).

Although there are membrane systems available for generating ozone at small scale, at waterworks scale it is invariably generated in an exothermic reaction by passing a silent corona discharge through air or oxygen:

$$3O_2 + hv \rightarrow 2O_3 + \text{heat} \quad \Delta H = -68 \text{ kcal mol}^{-1}$$

Typically the discharge is produced by an alternating power supply of 5–10 kV either in the range 50–60 Hz (low frequency) for small units or 100–1000 Hz (medium frequency) for larger installations. It is applied across a glass dielectric in the form of a tube with a metallised internal surface. The tube is mounted concentrically inside a stainless steel tube along which the gas flows and the assembly is surrounded by a cooling jacket. Having generated ozone it is important to ensure that the contacting system maximises the transfer of ozone into water. This is also important from the health and safety standpoint in that ozone which is not transferred to the water is vented off. Since ozone is highly toxic, the off-gas has to pass through an ozone destructor, either thermal (350–370°C) or catalytic (using a manganese, palladium or nickel oxide based catalyst). An example ozone contactor is shown in Figure 8.6. This one is designed to treat 1250 m^3 hr^{-1} at a 10.5 min contact time. Gas is passed

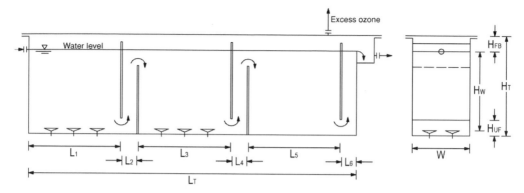

Fig. 8.6 Schematic of an ozone contactor (total length 12.2 m, width (W) 5 m, chamber length (L1) 3 m, water height above diffusers (Hw) 5 m).

directly to diffusers located at the bottom of the contact tank where bubbles are formed. These pass upwards through the water, and baffles are provided to create a counter-current flow. Ideally the water depth should be 5 m to allow for efficient gas transfer. The ozone dose is typically tapered with the higher dose being applied in the first stage. Typical transfer efficiencies for a well designed contacting system will be in the range 90–97%.

As with carbon the performance is very dependent on the molecules to be removed. Table 8.2 compares the first-order rate constants for the reaction between ozone and four molecules. Here it can be seen that microcystin will react rapidly with ozone, as does isoproturon. In comparison atrazine reacts slowly with ozone but as shown above it is readily removed by carbon adsorption, therefore, there would be little benefit to having ozone prior to GAC if atrazine was the target molecule. Ozone is unsuitable for the groundwater pollutant MTBE.

Table 8.2 Comparison of first-order rate constants of target molecules with ozone (adapted from Hall & Sheppard, 2005 & von Gunten, 2003).

Molecule	k_{O_3} $(M^{-1}s^{-1})$	Half life $(t_{1/2})^*$
Microcystin	3.4×10^4	1 s
Isoproturon	305	30 s
Endrin	<0.02	20 days
Atrazine	6	96 min
MIB	10	>1 hr
Geosmin	10	>1 hr
Estradiol	7×10^9	5 μs
MTBE	0.14	2.8 days

*Estimated for a 1 mg l^{-1} ozone dose.

References

Awwa (2000) Water Quality and Treatment. McGraw-Hill, New York, USA.

Brunauer, S., Emmett, P.H. & Teller, E. (1938). Adsorption of gases in multimolecular layers. *J. Am. Chem. Soc.*, **60**, 309.

Hall, T. & Sheppard, D. (2005). Ozone and GAC: striking the balance. CIWEM Scientific Group Seminar – GAC, pests and pesticides, Peterborough, 16th February.

Langmuir, I. (1918). The adsorption of gases on plane surfaces of glass, mica and platinum. *J. Am. Chem. Soc.*, **40**, 1361–1403.

von Gunten, U. (2003). Ozonation of drinking water: Part I. Disinfection and by-product formation in presence of bromide, iodide or chlorine. *Water Res.*, **37**, 1469–1487.

Disinfection 9

9.1 Introduction

The primary purpose of disinfecting water supplies is to inactivate microbial pathogens to prevent the spread of waterborne diseases. Waterborne diseases are typically caused by enteric pathogens which belong to the group of organisms transmitted by the faecal–oral route. These pathogens comprise a diverse group of organisms that serve as the agents of waterborne disease including bacterial, viral and protozoan species (Table 9.1). Disinfection of water supplies on a continuous basis was first attempted in England in 1904, and from 1912, with the development of the facilities for feeding gaseous chlorine, spread rapidly. This played a large role in the reduction of the death rate due to typhoid. While many of the other treatment processes described in this book can reduce the concentration of microbial pathogens, it is disinfection that serves as the final barrier to these organisms entering the potable water supply.

Disinfection is most commonly accomplished by the use of either:

1. Chemical agents (chlorine and its compounds, bromine, iodine, ozone, heavy metals and related compounds and hydrogen peroxide) or
2. Physical agents including heat, light or physical separation by microfiltration.

This chapter will focus on those disinfectants used for potable water production and cover the science, their efficacy and the by-products they form.

9.2 Process science

Two factors govern the effectiveness of the disinfection process: the contact time and the concentration of the disinfecting agent. The inactivation of

Table 9.1 Waterborne pathogens of concern (adapted from USEPA, 2001).

Organism	Size (μm)	Point of origin	Example	Resistance to disinfection	Removal by sedimentation and filtration
Bacteria	0.1–10	Humans, animals, water and contaminated food	*Salmonella* spp, *Shigella* spp, pathogenic *Escherichia coli*, *Campylobacter* spp, *Vibrio cholerae* and *Yersinia enterocolitica*	Type specific – bacteria spores have high resistance, vegetative spores have low resistance	2–3 log
Viruses	0.01–0.1	Humans, animals, polluted water and contaminated food	Hepatitis A and E, enteroviruses, adenoviruses, small round structured viruses including Norwalk virus, astro and rota viruses	Generally more resistant than vegetative bacteria	1–3 log
Protozoa	1–20	Humans, animals, sewage, decaying vegetation and water	*Entamoeba histolytica, Giardia intestinalis, Cryptosporidium parvum*	More resistant than viruses or vegetative bacteria	2–3 log

micro-organisms was found by Chick in 1908 to be characterised by a first-order rate law.

$$dN/dt = -kN$$

Where N is the number of viable organisms and k is the reaction rate constant dependent upon the disinfectant, the percentage kill required and the type of micro-organism. It was subsequently revealed that the rate constant, k, relates to the concentration of disinfectant, C:

$$k = k'C^n$$

Where n is the coefficient of dilution and k' is independent of disinfectant concentration. The Chick–Watson law may then be integrated to give the following relationship for batch reactions:

$$\ln \frac{N}{N_0} = -k'C^n t$$

This relationship assists the design of disinfection process systems and allows the comparison of different disinfectants to be easily made. Plotting combinations of disinfectant concentration and time to produce a fixed percentage inactivation is generally possible. Such plots tend to follow the relationship $C^n t = $ constant, where the constant is a function of the

Table 9.2 Reductions of bacteria, viruses and protozoa achieved by disinfection processes (adapted from WHO, 2004).

| | Ct for a 2 log reduction, mg min l^{-1} (pH ~7 unless stated) | | | | |
	Chlorine	Chloramines	Chlorine dioxide	Ozone	UV
					Ct for a 2 log inactivation, mJ cm^{-2}
Bacteria	0.08 at 1–2°C; 3.3 at 1–2°C pH 8.5	94 at 1–2°C; 278 at 1–2°C, pH 8.5	0.13 at 1–2°C; 0.19 at 1–2°C pH 8.5	0.02 at 5°C	7
Viruses	12 at 0–5°C; 8 at 10°C	1240 at 1°C; 430 at 15°C	8.4 at 1°C; 2.8 at 15°C	0.9 at 1°C; 0.3 at 15°C	59
Protozoa	*Giardia* 230 at 0.5°C; 100 at 10°C *Cryptosporidium* not killed	*Giardia* 2550 at 1°C; 1000 at 15°C *Cryptosporidium* not inactivated	*Giardia* 42 at 1°C; 7.3 at 25°C *Cryptosporidium* 40 at 22°C, pH 8	*Giardia* 1.9 at 1°C; 0.63 at 15°C *Cryptosporidium* 40 at 1°C; 4.4 at 22°C	*Giardia* 5 *Cryptosporidium* 10

organism, pH, temperature, form of disinfectant and extent of inactivation. Table 9.2 compares *Ct* values for bacteria, viruses and protozoa.

9.3   Chlorine

Chlorine is by far the most common oxidant used in water treatment (White, 1999). It is supplied either under pressure as a liquefied gas (Cl_2), or as sodium hypochlorite (bleach) containing about 13% available Cl_2 or as solid calcium hypochlorite (tropical bleach or bleaching powder) containing 30% available chlorine. The advantage of chlorine disinfectants is that they form a residual which remains in the water for long periods and protects it against bacterial contamination in the distribution system. Chlorine gas dissolves in water to form hypochlorous acid which further dissociates to form hypochlorite ions:

$$Cl_2 + H_2O \Rightarrow HCl + HOCl$$
$$HOCl \Rightarrow H^+ + OCl^-$$

At pH levels below 5 almost all the hypochlorous acid is undissociated whilst at pH levels above 9 dissociation is almost complete. Hypochlorous acid disinfects by penetrating the bacterial cell walls and attacking the cytoplasmic layer. In the unionised form it diffuses more readily through the cell wall than the hypochlorite (OCl^-) anion and therefore is a more efficient biocide, so the process benefits from maintaining a low pH

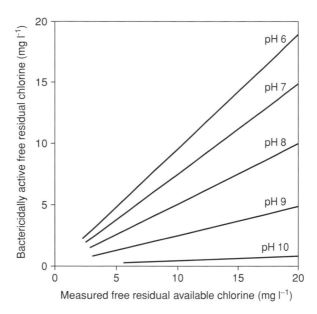

Fig. 9.1 Relationship between measured free residual available chlorine (HOCl, OCl⁻) and bactericidally active (HOCl) (WHO, 2004).

(Figure 9.1). Unfortunately, at these low pH values (less than 6) degradation of HOCl to oxygen and hydrochloric acid (HCl) becomes significant over the disinfection contact time period of 30 min or more, this time being required for effective disinfection. There is therefore a trade off between promoting efficient inactivation of micro-organisms and suppressing the breakdown of chlorine.

In potable water treatment, the pH is to a large extent determined by other factors, including legal stipulations and preceding unit processes. However, the stability of chlorine is such that dosing to around 0.5–1.0 mg l⁻¹ after treatment at the water treatment works is usually sufficient to leave a residual level of 0.1–0.2 mg l⁻¹ chlorine (as Cl_2, HOCl or OCl⁻) at the point of use. This level is normally taken as being indicative of the water being safe to drink without imparting a chlorinous taste to the water.

The dosage of chlorine required is dependent on the water source, the organism and the required kill rate. The chlorine residual and contact times vary depending on the species of vegetative bacteria, viruses and amoebic cysts present (Table 9.3) and generally a safety factor is applied to take into account a more resistant species. The WHO recommends the following conditions for disinfections with chlorine which equates to a *Ct* of 15 mg min l⁻¹. Typical *Ct* values for UK waters are shown below

Table 9.3 Summary of *Ct* values (mg min l^{-1}) for 99% inactivation at 5°C (Clark *et al.*, 1993).

Micro-organism	*Ct* for 2 log reduction, pH 6–7
E. coli	0.034–0.05
Polio 1	1.1–2.5
Rota Virus	0.01–0.05
Phage f2	0.08–0.18
G. lamblia cysts	47–>150
G. muris cysts	30–630
C. parvum oocysts	7200

(Table 9.4) and are in line with WHO guidelines although there are lower *Ct*s for groundwaters where there is little evidence of contamination.

$$\text{Residual free chlorine} \geq 0.5 \text{ mg } l^{-1}$$

$$\text{Contact time} \geq 30 \text{ min}$$

$$\text{pH} < 8.0$$

$$\text{Maximum turbidity} = 5 \text{ NTU}$$

Whether the chlorination medium is chlorine gas, sodium hypochlorite stock solution or on-site electrolytically generated hypochlorite, it is dosed at the inlet to a *contact tank* (Figure 9.2). This is designed as a plug flow reactor with residence time typically 20–60 min. The example shown below is designed to treat 635 m^3 hr^{-1} at a 30 min contact time. The water leaving the contact tank will usually require a chlorine residual of about 0.2 mg l^{-1} and, because of the loop time between the dosing point and measurement point it is a common practice to dose excess chlorine at the reactor inlet and to de-chlorinate by dosing sulphur dioxide or sodium bisulphite at the outlet. The de-chlorination reaction is very rapid and is simple to control.

9.4   Chloramination

Chloramines can be generated by deliberately adding ammonia (in the form of ammonium sulphate) simultaneously with chlorine, a process

Table 9.4 Typical *Ct* values for UK waters.

Source	Quality	*Ct* (mg min l^{-1})
Groundwater	No *E. coli*	5
Groundwater	*E. coli*	15
Surface water	Lowland	20
Surface water	Moorland	30

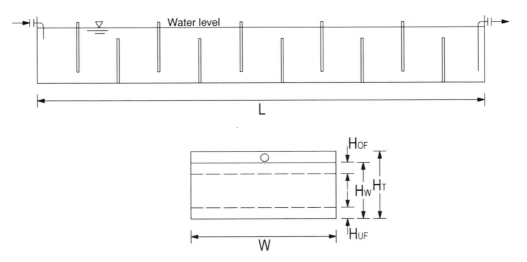

Fig. 9.2 General arrangement of disinfection contact tank (where length (L) 20 m, width (W) 6.5 m, water depth (Hw) 2.5 m, total height (HT) 3 m, baffle underflow height (HUF) 0.5 m, baffle overflow height (HOF) 0.5 m, baffle height 2 m).

called chloramination. Chloramines are less active disinfectants than chlorine but are very stable and are often used in distribution systems which have a large capacity and long retention time. Chloramination is becoming popular as chloramines are much weaker oxidising agents than either hypochlorous acid or hypochlorite.

The reaction between ammonia and chlorine initially generates monochloramine then dichloramine and trichloramine which are known collectively as *combined chlorine*:

$$NH_4^+ + HOCl \Leftrightarrow NH_2Cl + H_2O + H^+$$
$$NH_2Cl + HOCl \Leftrightarrow NHCl_2 + H_2O$$
$$NHCl_2 + HOCl \Leftrightarrow NCl_3 + H_2O$$

The rate formation is also pH dependent and chloramination is normally carried out using a chlorine:ammonia (as N) ratio of 5 : 1 by weight, which minimises both the free chlorine residual and the free ammonia concentration although for some waters, the required chlorine:ammonia molar ratio may be as high as 7 : 2 (a mass ratio of 15 : 1 weight ratio of Cl_2 to NH_3) but is usually around 2 : 1. The presence of free ammonia in the distribution system can give rise to the growth of nitrifying bacteria (*Nitrosomonas*) which can result in an increase in nitrite concentration in the water with attendant health risks. In most modern applications ammonia is dosed after chlorination and after a short contact time.

9.5   Ozone

Ozone (O_3) is produced by passing a high voltage electrical corona discharge through dry air or oxygen. It is a more powerful oxidant than hypochlorous acid and is approximately 5 times more effective as a disinfectant (see Chapter 8). Ozone has been a popular treatment process with the UK water utilities. A survey of six of the UK utility companies showed 31 water treatment works used ozone ranging from a few MLD through to a 360 MLD plant. Most of these plants though have been primarily installed for the destruction of refractory organic compounds, and pesticides in particular. Although much less soluble than chlorine, by about 20 times, it requires a shorter contact time for deactivating most micro-organisms – about 10–12 min is normally sufficient.

9.6   Chlorine dioxide

Chlorine dioxide can be used as a primary or secondary disinfectant and it is used for controlling taste and odour as well as zebra mussel control. Chlorine dioxide rapidly inactivates most micro-organisms over a wide pH range and it is typically more effective than chlorine but is less effective than ozone. Chlorine dioxide is a powerful oxidant and will react with a wide range of organic and inorganic compounds.

9.7   Ultraviolet light

There is a growing interest in the use of ultraviolet (UV) light to disinfect drinking water as it has been shown to inactivate a wide range of micro-organisms including *Cryptosporidium* (see Table 9.2) without producing any disinfection by-products. UV light is widely used in the UK to disinfect wastewater discharges at coastal sites but its use in drinking water applications was primarily limited to small groundwater systems.

Ultraviolet radiation is an electromagnetic radiation of slightly higher frequency than visible light but rather lower than that of X-rays. It is responsible for the tanning effect of sunlight and, it is argued, is a major cause of malignant melanoma or skin cancer. The UV spectrum is arbitrarily divided into three bands according to the wavelength of the radiation It is the lowest wavelength (and therefore highest frequency) radiation, in the UVC band, that has the strongest biocidal properties.

UVA	400–315 nm
UVB	315–280 nm
UVC	280–200 nm
Visible	400–700 nm

Table 9.5 UV lamp characteristics.

Parameter	Low pressure	Medium pressure	High pressure
UV wavelength (nm)	185 and 254	240–300	240–300
Max output (W m^{-2})	60	500–2000	>2500
Typical lamp life (hr)	8000	4000	4000
Optimum temp (°C)	50*	0–100	0–100

*Output from a low-pressure lamp falls off as the surface temperature of the lamp varies from 50°C in either direction.

UV radiation is produced commercially by the use of mercury vapour, antimony and xenon lamps. Normally the lamp is enclosed in a protective quartz sleeve and its output is expressed as the UV radiation power measured at the outer surface of the sleeve in W m^{-2} of surface area. Commercial lamps fall into two principal categories: low pressure and medium pressure. Their characteristics are summarised in Table 9.5 and output shown in Figure 9.3.

Ultraviolet irradiation owes its bactericidal effect to its ability to penetrate the cell and act directly on the nuclear DNA. The radiation does not destroy the bacterial cell material, but disrupts the DNA by causing adjacent chemical groups on the double helix of the DNA molecule to fuse and prevent the molecule from replicating. This means that the bacterium is unable to reproduce and is thus *inactivated* or *not viable*. It is, however, wrong to think in terms of bacteria being killed by UV because it has been demonstrated that exposure of a UV-treated bacterial cell to visible light (300–500 nm) causes *photo-reactivation* which reverses the effects of the UV dose. UV dose is often measured in millijoules per square centimetre or milliwatts per square centimetre (10 J m^{-2} = 1 mJ cm^{-2} = 1 mW cm^{-2})

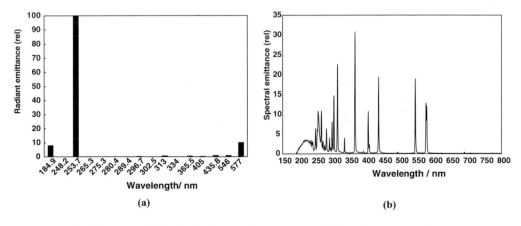

Fig. 9.3 Spectral distribution from (a) low-pressure and (b) medium-pressure lamps.

Table 9.6 Reduction in *E. coli* at increasing UV dose.

UV dose, mJ cm^{-2}	Reduction in viable count (%)
5.4	90
10.8	99
16.2	99.9
21.6	99.99

and the relationship between UV dose and inactivation rate is logarithmic as Table 9.6 shows for *E. coli*.

The understanding of UV technology has developed enormously in the past 5 years due to the application of computational fluid dynamics to optimise UV reactor design and by challenging UV reactors with resistant microbes to demonstrate performance. This has built a confidence in UV reactors' ability to inactivate microbes under a range of realistic water treatment plant operating conditions; i.e. water quality, flow rate and turbidity. A wide range of UV reactors have been designed to efficiently deliver the necessary dose to inactivate micro-organisms (see Figure 9.4). The USEPA has accepted that UV reactors, which have been independently validated, can be used for *Cryptosporidium* protection. The USEPA has published comprehensive guidelines on validation and the issues to be addressed in UV reactor design. This means that UV reactors offered

Fig. 9.4 UV installation at Victoria, British Columbia (with kind permission from Trojan Technologies).

Table 9.7 Examples of potable water treatment works where UV is being used for *Cryptosporidium* inactivation (data courtesy of Trojan UK).

	Flow (m^3 hr^{-1})	Water source	Bioassay dose (mJ cm^{-2})	Performance (inactivation)
Seattle, Washington	28 000	Surface water	40	3 log *Cryptosporidium*
Victoria, British Columbia, Canada	24 000	Surface water	40	2 log *Cryptosporidium*
Rotterdam, The Netherlands	28 400	Surface water	70	1.5 log clostridia + *Cryptosporidium*
Albany, New York	6300	Surface water	40	2 log *Cryptosporidium*

commercially should be able to demonstrate the ability to deliver a bioassay dose which is internationally accepted. There are other internationally recognised standards for bioassay validation like the DVGW in Germany and O-NORM in Austria.

UV will effectively disinfect waterborne pathogens with a UV dose of 40 mJ cm^{-2} and achieve a 4 log reduction, an exception is the adenovirus which with a UV dose of 40 mJ cm^{-2} gives a 1.5 log inactivation, but adenovirus is highly susceptible to a low chlorine dose. A chlorine residual should still be used for protection in the distribution system and there is growing interest (see Table 9.7) in a multi-barrier approach to disinfection of municipal drinking water, with UV as the primary disinfectant and a chlorine residual providing protection in the distribution system. The major factors affecting the performance of the UV process include (i) UV transmission, (ii) turbidity, (iii) hydraulics and (iv) foulants such as iron, organic matter and calcium carbonate. The transmissivity of the water is of crucial importance in determining how much UV power must be applied to ensure that all the water in the cell is exposed to the desired dose; and the flow rate, which a given lamp unit will treat, is directly proportional to the transmissivity of the water. Light transmission is affected by a number of characteristics of the water. Colour in the water absorbs UV and visible light, iron salts are oxidised and 'use up' UV radiation, whilst turbidity scatters the light. Suspended solids in the water may shield micro-organisms from the radiation and may also coat the quartz sleeve, reducing the intensity of UV radiation reaching the water. Scale deposition has the same effect. Solids are particularly significant when virus removal is required, since viruses are associated mostly with the suspended solids.

9.8 Disinfection by-products

Disinfection by-products (DBPs) are formed when chemical disinfectants react with natural organic matter and inorganic ions such as bromide and

Table 9.8 Major DBPs formed by different disinfection processes.

Disinfectant	Inorganic by-product or disinfectant residual formed
Chlorine	Trihalomethanes, haloacetic acids, haloacetonitriles, haloketones, chlorophenols, MX
Chlorine dioxide	Chlorine dioxide, chlorite, chlorate, bromate
Ozone	Bromate, iodate, hydrogen peroxide
Chloramines	Monochloramine, dichloramine, trichloramine, ammonia, cyanogen chloride

iodine present in water (Table 9.8). These by-products are varied and can include the trihalomethanes (THMs) and haloacetic acids (HAAs). These two groups of DBPs in particular have led to concern by the regulators such as the USEPA and the WHO, as many have been shown to cause cancer in animals (Singer, 1999). Legislation has tightened to control the amount of DBPs allowed in drinking water and currently the UK allows an upper limit of 100 μg l^{-1} for total THMs (trichloromethane, dichloro-bromomethane, dibromochloromethane and tribromomethane) based on a spot sample, whilst US regulations allows a maximum of 80 μg l^{-1}. The US has also imposed a limit on the total of five HAAs (monochloro-, dichloro-, trichloro-, monobromo- and dibromo-acetic acids) of 60 μg l^{-1}. Whilst THMs and HAAs are often the major DBPs (see Table 9.8) formed, recent studies have identified over 500 different DBPs and which one is formed will be very dependent on the water quality and the disinfectant used. The chemical structures of the major groups of DBPs are shown in Figure 9.5.

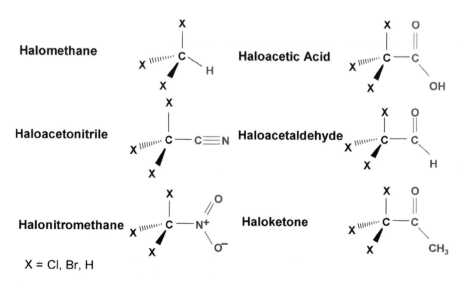

Halomethane **Haloacetic Acid**

Haloacetonitrile **Haloacetaldehyde**

Halonitromethane **Haloketone**

X = Cl, Br, H

Fig. 9.5 Chemical structures of major DBPs.

References

Clark, R.M. & Regli, S. (1993) Development of Giardia Ct values for the surface water treatment rule. *J. Environ. Sci. Health*, **28**, 1081–1097.

Singer P.C. (1999). Humic substances as precursors for potentially harmful disinfection by-products. *Water Sci. Technol.*, **40**, 25–30.

White, G.C. (1999). *Handbook of Chlorination and Alternative Disinfectants*. Wiley, New York.

WHO (2004) *Guidelines for Drinking-water Quality* (2004). World Health Organization, Geneva.

USEPA (2001) Controlling Disinfection By-Products and Microbial Contaminants in Drinking Water. EPA/600/R-01/110, Washington DC, USA.

Organics Removal 10

10.1 Introduction

An organic compound has a chemical structure based on the carbon atom. There are millions of natural or synthetic organic compounds containing carbon, including hydrocarbons, sugars, proteins, plastics, petroleum-based compounds, solvents, pesticides and herbicides. Many of these compounds have been identified in surface and ground waters and typically originate from the following sources:

1. Decomposition of naturally occurring organic materials in the environment;
2. Industrial, agricultural and domestic activities; and
3. Reactions occurring during the treatment and distribution of drinking water.

Water used for potable supply will, depending on its source, contain natural and synthetic organic compounds. Moorland and upland source waters have high concentrations of natural organic matter whilst lowland waters are likely to have inputs from both agricultural and anthropogenic sources. Table 10.1 gives an overview of the performance of a range of conventional and advanced water treatment processes for removing them from water.

In this chapter we are going to focus on four main groups of organics and how water treatment works are adapted to remove them. The four groups are listed below:

- Disinfection by-products precursors
- Pesticides and other micro-pollutants such as endocrine disrupting chemicals and pharmaceuticals
- Algae
- Taste and odour.

Table 10.1 Comparison of organics removal in conventional and advanced water treatment processes.

	Coag/Floc	IEX	GAC	O₃/GAC	Membrane	AOPs
NOM	√√√	√√	√	√	√√	√√
Pesticides	√		√√	√√√	√√	√√√
Pharmaceuticals	√		√	√√		√√
MTBE			√			√√
Taste and odour			√√	√√√	√	√√
Algae	√√√		√	√	√√	√

√√√ – excellent removal, √ – poor removal.

10.2 Disinfection by-products

Most water sources throughout the world contain dissolved organic matter, often termed natural organic matter (NOM). NOM is best described as a complex mixture of organic compounds and has been shown to consist of organics as diverse as humic acids, hydrophilic acids, proteins, lipids, hydrocarbons and amino acids. The range of organic components varies from water to water as well as seasonally. NOM itself is considered harmless, however, legislation requires that disinfection is applied to drinking water in order that the water remains fit for human consumption when it reaches the tap. It is the conversion of NOM into disinfection by-products (DBPs) when chlorine is used that can cause problems. The DBPs are discussed in detail in Chapter 9. They can be in the form of trihalomethanes (THMs), haloacetic acids (HAAs) and many other halogenated compounds, some of which are of major concern to the water treatment industry, as tests have shown links between cancer in laboratory animals and THMs. To control the risk US and UK legislation has tightened in recent years to dictate the amount of THMs allowed in drinking water. The current consent in the UK is 100 μg l⁻¹ for total THMs (TTHMs) and 80 μg l⁻¹ in the US (see Chapters 1 and 9).

In the UK organics levels have increased significantly in waters drawn from moorland and upland catchments, which has impacted on both colour levels and the amount of THMs formed in distribution, see Figure 10.1. The example given below is for a water treatment works in Halifax, UK where colour levels have increased fourfold over the past 15 years leading to increased coagulant use. Here optimisation of the treatment works and specifically the coagulation process has minimised the impact on THMs measured in the distribution system.

The treatment of water for potable use has traditionally focussed on the removal of either colour or turbidity, but it is clear that to control THM and HAA formation we must reduce the concentration of precursors and

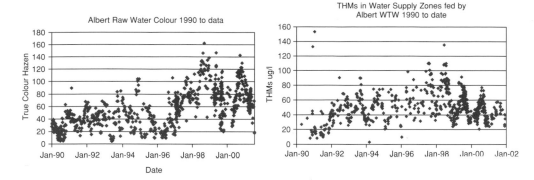

Fig. 10.1 Example of annual changes in colour and THM levels as a result of elevated NOM levels (adapted from Banks & Wilson, 2002).

specifically reduce the overall dissolved organic carbon (DOC) content of water. There are a number of proven technologies for the removal of NOM used in the water treatment industry. These include physical, chemical and biological processes and are each capable of different DOC removal performance (Table 10.2). This section will focus on coagulation, ion exchange and advanced oxidation process options for removing DBP precursors.

10.2.1 Coagulation

As NOM is almost always anionic at the pH of natural water it has a strong affinity to cationic additives such as metal coagulants and cationic polyelectrolytes. Consequently, coagulation is regarded as the best available process when removing dissolved organic matter. Coagulation with iron or aluminium salts is specifically good at removing hydrophobic and high molecular weight organic matter such as humic and fulvic acids. It is not so good at removing the uncharged and hydrophilic organic matter such as alkyls and polysaccharides. UV_{254} and a parameter called specific

Table 10.2 Summary of process options and performance for DOC removal.

Process	Typical DOC removal (%)	Disadvantages
Coagulation	10–60	High coagulant doses and sludge production. Removal efficiencies related to water source
Ion exchange/adsorption	60–80	Higher costs. Generation of waste streams
Membranes	80–100	High costs. Still requires pre-treatment to prevent membrane fouling
Ozonation/biodegradation	27–75	Variable removal efficiencies
Advanced oxidation processes	60–90	Not proven at full scale

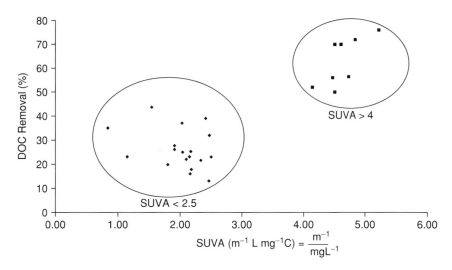

Fig. 10.2 Relationship between SUVA and DOC removal by coagulation.

UV absorbance (SUVA, which is the ratio of UV absorbance at 254 nm in m^{-1} to DOC concentration in mg l^{-1}) both give relative indications of the hydrophobic content in water. They give reasonable predictions of the formation of THMs and are good predictors of the performance of coagulation; whereas SUVA increases so does the removal of dissolved organic carbon (Figure 10.2). If a water is hydrophobic in nature (SUVA > 4) 50% removal of DOC will normally be achieved, but if the water is hydrophilic in nature (SUVA < 2.5) much lower removal efficiencies will be achieved. High SUVA waters tend to be from upland or moorland catchments whilst lowland sources and groundwater have SUVA values of less than 2.5.

10.2.2 Ion exchange

One process that has emerged in the past few years is the magnetic ion exchange or MIEX® DOC process. The process was developed by Orica Watercare, South Australian Water Corporation and the CSIRO and is an ion exchange process specifically for the removal of DOC from drinking water. The resin is a strong base anion exchange resin with a macroporous structure and type 1 quaternary ammonium active sites attached to a magnetic core. The process involves adsorbing the DOC onto the MIEX® resin in a stirred contactor that disperses the resin beads to allow for maximum surface area. The magnetic part of the resin allows the resin to agglomerate into larger, faster settling particles which allow for a recovery rate of greater than 99.9%. Any resin that is carried over is removed in downstream processes, either by filtration or micro-filtration (Figure 10.3). When combined with a low dose of coagulant the MIEX®

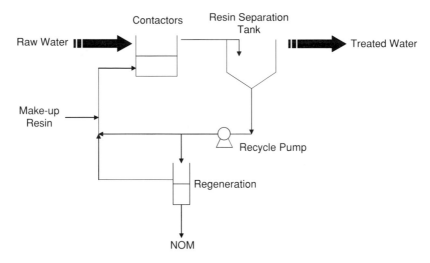

Fig. 10.3 Schematic of MIEX® process.

process is able to match the DOC removals achieved by coagulation and removes more of the hydrophilic low molecular weight organics. This can lead to lower THMs than just coagulation alone with reduced coagulant dose and also sludge production (Figure 10.4).

10.2.3　Advanced oxidation processes

A range of advanced oxidation processes (AOPs) have been tested for treating water containing commercial humic acid or NOM. AOPs work by producing a reactive species called a hydroxyl radical in water. The

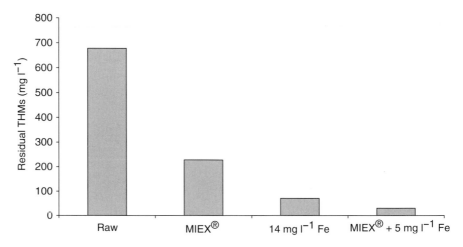

Fig. 10.4 Comparison of conventional coagulation, MIEX and MIEX plus coagulation in controlling THMs (adapted from Fearing et al., 2004).

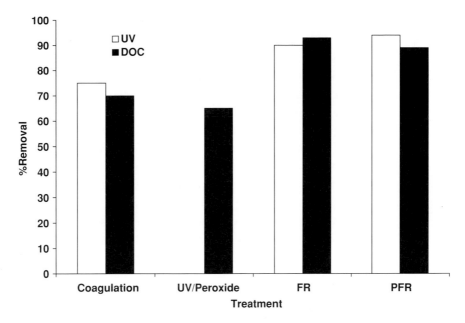

Fig. 10.5 Maximum % UV_{254} and DOC removal from raw water with Fenton's, photo Fenton's, H_2O_2/UV oxidation and conventional coagulation (adapted from Murray & Parsons, 2004).

hydroxyl radical is able to oxidise a range of organic compounds significantly (10^9) faster than ozone, for example the rate constant for the reaction between hydroxyl radicals (OH) and humic acid is 3×10^4 mg l^{-1} DOC s^{-1}, and is significantly faster than ozone (Murray & Parsons, 2004). A comparison of three AOPs, UV/H_2O_2, Fenton's and photo-Fenton's, for treating an upland catchment's reservoir water (DOC 7.5 mg l^{-1}) showed how both Fenton's processes could remove in excess of 90% of DOC and UV adsorbing species (Figure 10.5). This was compared to the ~70% achieved by conventional coagulation. Both Fenton's processes remove in excess of 70% of raw water UV_{254} after 1 min, with both achieving a maximum removal of 95% PFR after 5 min, compared to 30 min for FR, significantly quicker than UV/H_2O_2.

10.3 Micro-pollutants

Whilst there are obvious issues regarding the reaction of organics with chlorine, there are a number of other questions relating to what organics are present in the water passing through conventional water treatment processes. The material is often polar in nature and is a heterogeneous mixture of organic molecules characterised by high solubility in water; examples of chemical groups identified in this fraction include alcohols, ketones, esters, alkylhalides and aromatics. Few individual compounds have been identified but any organic compound with a high water solubility (often characterised by low K_{ow} value) has the potential to be in this fraction. A review of the literature has shown that a number of current and emerging pollutants fit into the same group: (i) a third of all pharmaceutical products previously identified in surface waters, (ii) four of the most popular 12 pesticide products used on arable crops in the UK and (iii) a number of emerging pollutants such as MTBE, vinyl chloride and haloacetic acids. These hydrophilic organics are poorly removed during wastewater treatment and can enter surface and ground waters from wastewater discharges and then on into potable supply.

10.3.1   Pesticides

Pesticides can be classified according to their application either as an insecticide, herbicide or fungicide. The use of pesticides has developed enormously since the 1950s. They are particularly hazardous as they are chemically developed to be toxic and to some extent persistent in the environment. In permeable soils they are able to infiltrate to groundwater and in more impermeable soils they find their way into surface watercourses via drainage networks. Pesticide levels in drinking waters taken from rivers tend to be a seasonal phenomenon, although they can be found at relatively high concentrations throughout the year. In groundwaters they accumulate over the years so that the overall concentration is slowly but constantly increasing, without any discernible large seasonal variation. In Europe, the Drinking Water Directive sets a Maximum Allowable Concentration (MAC) value for total pesticides and any individual pesticide compound of 0.5 and 0.1 ppb, respectively (see Chapter 1). Total pesticides is expressed as 'pesticides and pesticide related products', which is defined as insecticides (persistent organochlorine compounds and carbamates), herbicides, fungicides, polychlorinated biphenyls (PCBs) and polychlorinated terphenyls (PCTs).

The best available technology for removing pesticides is activated carbon. The adsorption capacity of activated carbon to remove pesticides is affected by concentration and physical/chemical properties of the contaminant as well as loading rate and empty bed time. Generally, activated carbon has an affinity for contaminants that are hydrophobic (low solubilities), so any pesticides exhibiting a high organic adsorption coefficient (K_{oc}) and low solubility are expected to exhibit high binding affinities for

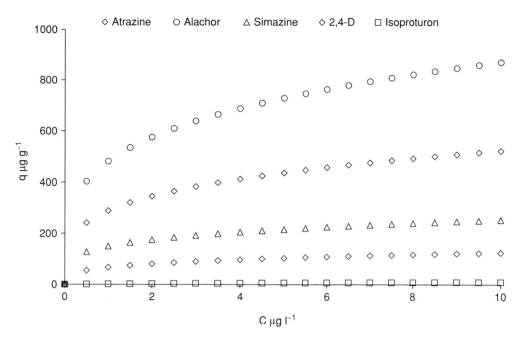

Fig. 10.6 Comparison of predicted Freundlich isotherms for a range of pesticides.

activated carbon. In general, compounds with Freundlich coefficients on activated carbon greater than 200 $\mu g\ g^{-1}$ $(1\ \mu g^{-1})^{1/n}$ would be amenable to removal by carbon sorption (Speth & Miltner, 1998). Figure 10.6 compares the predicted performance of activated carbon for removing a range of pesticides. The figure was produced using literature values of Freundlich coefficients (K_f and n) and shows that carbon is more suitable for alachor, atrazine and simazine, but would probably need ozone as a pre-treatment if the removal of isoproturon is required.

AOPs have also been proposed for removing pesticides and in particular using UV and a combination of UV and H_2O_2. Richardson and Holden (2004) have reported data from pilot and field trials of UV and also a combination of UV and H_2O_2, and found the process to be competitive with GAC for a range of pesticides (Figure 10.7). The results indicate mecoprop is readily degraded by UV photolysis and 86% removal was achieved with a UV dose of 2000 mJ cm^{-2} for surface waters with a UV transmittance (UVT) of 80%. At the same dose the triazine group of pesticides can be degraded by 50% and diuron and isoproturon by 30 and 19%, respectively. It is worth comparing the significantly higher UV dose required for photolysis to disinfection (see Table 9.2). Along with dose the UVT is a key parameter in the performance of any UV process

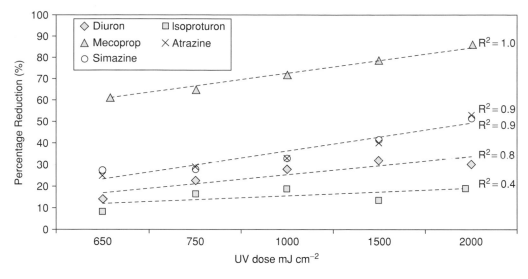

Fig. 10.7 Percentage removal of pesticides at each UV dose studied (adapted from Richardson & Holden, 2004).

and the higher the UVT the better the process will perform. For example groundwater quality is typically higher than surface water and Richardson and Holden (2004) showed that when UVTs were greater than 95% it was possible to achieve 95% removal for mecoprop.

10.3.2   Endocrine disrupting compounds

There is increasing concern about the environmental effects of oestrogens and xenoestrogens in water (USEPA, 2001a). Oestrogenic substances act as endocrine disruptors and reported effects include the perturbation of sexual differentiation in embryos, cancers in reproductive organs and altered glucose and fat metabolism (Parsons, 2004). The three main human oestrogen compounds are 17-β-oestradiol (Figure 10.8), oestriol and oestrone, of which the former is the most potent. Oestrogens may enter the aquatic environment from contraceptive pill residues, hormone replacement therapy residues and diethylstillbestrol residues (used to promote livestock growth). Also cause for concern are xenoestrogen substances, a wide range of diverse compounds which mimic the biological effects of oestrogens. Substances classed as xenoestrogens include specific pesticides, e.g. DDT analogues and PCBs, and also alkylphenol polyethoxylates which are surfactants used in detergents. Although not as potent as the oestrogens, xenoestrogens are more abundant in the aquatic environment. Normal processes of water and wastewater treatment are not completely effective in removing oestrogenic substances.

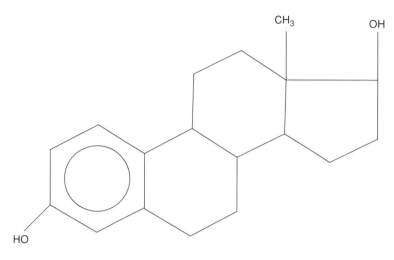

Fig. 10.8 Structure of 17-β-oestradiol.

Table 10.3 compares the performance in removing micro-pollutants including EDCs with conventional treatment processes such as coagulation/flocculation and activated carbon with AOPs and UV processes. It is clear that both UV and AOPs have great potential in treating the whole range of chemicals listed, typically better than coagulation with metal salts and comparable with activated carbon.

10.4 Algae

Algae are ubiquitous in surface water but do not pose a problem to water treatment processes provided populations are relatively low. In general, problematic algae are colloidal in nature and as such are removed by conventional processes, for example coagulation/flocculation followed by

Table 10.3 Unit processes and operations used for micro-pollutant removal (adapted from Parsons, 2004).

Classification	GAC	AOP	UV	Cl_2/ClO_2	Coag
Steroids	>90	>90	>90	<20	<20
Industrial chemicals	>90	>90	>90	>90	<20–40
Antibiotics	40–90	20–90	40–90	20–90	<20–40
Antidepressants	70–90	20–90	40–90	20–70	<20–40
Anti-infalamants	>90	20–90	70–90	20–70	<20
Lipid regulators	>90		>90	20–70	<20
X-ray contrast media	70–90	70–90	20–90	20–70	<20–40
Synth. musks	70–90	70–90	>90	20–70	<20–40
Antimicrobials	70–90	70–90	40–90	20–70	<20–40
Surfactants	>90	>90	40–90	<20	<20–40

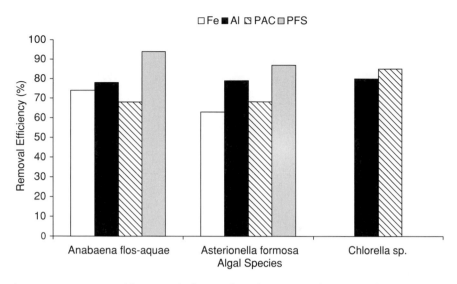

Fig. 10.9 Comparison of the removal efficiency by sedimentation of *Anabaena flos-aquae* (blue-green algae), *Asterionella Formosa* (diatom) (Jiang *et al.*, 1993) and *Chlorella* sp. (green algae) (Liu *et al.*, 1999) using four common coagulants (ferric, aluminium, polyaluminium chloride and polyferric sulphate).

either sedimentation or direct filtration. DAF is becoming more popular, as sedimentation is commonly regarded as an inefficient process due to the low density of algae. More novel processes are being investigated and implemented due to the variable characteristics of algae, including variations of the DAF discussed in Chapter 5. However, in most instances, coagulation/flocculation is the process that algae encounter initially. Algae are removed with varying efficiency dependent on coagulant type and dose, operational pH and subsequent separation process. Figure 10.9 compares the performance of four coagulants (Al – aluminium sulphate; Fe – ferric sulphate; PAC – polyaluminium chloride; PFS – polyferric sulphate) for three different algae species and shows how coagulation can achieve 70–90% removal.

However, continuing eutrophic conditions cause seasonal algal blooms and populations can increase dramatically, to the extent that they interfere with various unit operations if processes are not sufficiently optimised. Coagulation and flocculation can be impaired resulting in (a) high carryover of algae, (b) increased coagulant demand and (c) an increase in coagulant residual in the treated water. Filter clogging often occurs when large algae, for example diatoms, are present, which significantly reduces running times, while minimal retention of micro or motile algae is observed. Prechlorination to improve coagulation becomes unfeasible as algal cells and associated algogenic material are THM precursors. Furthermore, offensive taste and odour of the resultant drinking water supply is sometimes attributed to the presence of algae, and certain taxa have the potential for toxin release by some algal cells. In fact, research into toxin release from a particular species, *Microcystis*, has resulted in the WHO setting a guideline value of 1 μg l^{-1} for the associated toxin, microcystin-LR (Chapter 1). The WHO have since outlined risk levels whereby the presence of 100 000 cells ml^{-1} of *Microcystis* or the equivalent 50 μg l^{-1} as chlorophyll *a* in 4 m depth of water represents a moderate risk, producing in the region of 20 μg l^{-1} as microcystin-LR.

10.5   Taste and odour

Consumers will judge the quality of the water they drink based mainly on aesthetic properties such as taste and odour, and the presence of off-flavours in the water are linked with potential health risks (Suffet *et al.*, 1995; Suffet *et al.*, 1999). Two compounds that are commonly identified as causing taste and odour problems are MIB and geosmin. These are earthy, musty odour compounds produced as secondary metabolites by some cyanobacteria and actinomycetes. They can cause significant problems for drinking water suppliers as they are perceived by most consumers as an unpleasant odour and taste in water at very low concentrations (around 10 ng l^{-1}). Conventional water treatments such as coagulation, filtration and oxidation have been shown to be inadequate for the removal of MIB and geosmin. The compounds, aliphatic, tertiary alcohols, are not easily oxidised by chlorine or ozone, consequently activated carbon is currently considered the best technology for the treatment of affected water.

References

Banks, J. & Wilson, D. (2002) Use of UV$_{254}$ to predict the relationship between NOM and THMs on Upland Waters. *In International Conference on Natural Organic Matter Characterisation and Treatment*, Cranfield University, Cranfield, England, 15th May.

Fearing, D.A., Goslan, E.H., Banks, J., Wilson, D., Hillis, P.H., Campbell, A.T. & Parsons, S.A. (2004) Combination of ferric and MIEX® for the treatment of humic rich water. *Water Res.*, 38, 2551–2558.

Jiang, J., Graham, N.J.D. & Harward, C. (1993) Comparison of polyferric sulphate with other coagulants for the removal of algae-derived organic matter. *Water Sci. Technol.*, 27, 221–230.

Liu, J.C., Chen, Y.M. & Ju, Y. (1999) Separation of algal cells from water by column flotation. *Sep. Sci. Technol.*, 34, 2259–2272.

Murray, C.A. & Parsons, S.A. (2004) Comparison of AOPs for the removal of NOM: performance and economic assessment. *Water Sci. Technol.*, 49, 267–272.

Parsons, S.A. (ed.). (2004) *Advanced Oxidation Processes for Water and Wastewater Treatment*. IWA Publishing, London.

Richardson, A. & Holden, B. (2004) The evaluation of UV and UV/H_2O_2 for pesticide and algal toxin removal. *2nd International Conference on Advanced Oxidation Processes for Water and Wastewater Treatment,* Cranfield, 7 April, UK.

Speth, T. & Miltner, R. (1998) Technical note: adsorption capacity of GAC for synthetic organics. *J. Am. Water Works Assoc.,* **82,** 72–75

Suffet, I.H., Khiari, D. & Bruchet, A. (1999) The drinking water taste and odour wheel for the millennium: beyond geosmin and methyl isoborneol. *Wat. Sci. Technol.,* **40**(6), 1–13.

Suffet, I.H., Mallevialle, J. & Kawczyski, E. (ed.) (1995) *Advances in Taste-and-Odour Treatment and Control.* AWWA Research Foundation and Lyonnaise Des Eaux Cooperative Research Report, American Water Works Assoc., Denver, Colorado, USA.

USEPA (2001a) EPA/625/R-00/015. Removal of endocrine disruptor chemicals using drinking water treatment processes. U.S. Environmental Protection Agency, Cincinnati, Ohio, USA.

Inorganics Removal 11

11.1 Introduction

An inorganic compound is a chemical compound not containing carbon. However, elemental carbon (diamond or graphite) as well as carbon monoxide, carbon dioxide and carbonates are typically considered inorganic, while methane, ethanol and similar simple hydrocarbons are referred to as organic compounds. Inorganic contaminants are found in all water sources and can be present as cations, anions or neutral species and their occurrence depends primarily on their aquatic chemistry. Inorganics get into water through many routes including the dissolution of naturally occurring minerals, from industrial wastes and effluents, and from distribution and plumbing systems. Table 11.1 gives an overview of the performance of a range of conventional and advanced water treatment processes for removing inorganics from water.

In this chapter we are going to focus on the main regulated inorganic contaminants including nitrate, arsenic, fluoride, iron, manganese, bromate and lead. There are a number of other metal ions that are regulated, which are not covered here but many of the physicochemical processes described in this chapter are also suitable for these.

11.2 Nitrate

Nitrate levels in many UK waters, both ground and surface waters, are increasing and recently 55% of England has been designated as being in a nitrate vulnerable zonel (NVZ). An NVZ has been identified as having nitrate polluted waters using the following specific criteria:

- Surface freshwaters which contain or could contain, if preventative action is not taken, nitrate concentrations greater than 50 mg l^{-1}.

Table 11.1 Comparison of inorganic ion removal in conventional and advanced water treatment processes.

	Coag/floc	IEX	Oxidants	Reductants	Adsorbents	Membranes
Nitrate		√√		√√	√	√√√
Bromate	√	√√		√√	√	√
Arsenic	√√	√√	√		√√√	√√√
Fluoride	√√				√√√	√√
Iron	√	√	√√			√√
Manganese			√√			√√
Lead				√√	√√	

√√√ – excellent removal, √ – poor removal.

- Groundwaters which contain or could contain, if preventative action is not taken, nitrate concentrations greater than 50 mg l^{-1}.
- Natural freshwater lakes, or other freshwater bodies, estuaries, coastal waters and marine waters which are eutrophic or may become so in the near future if protective action is not taken.

Approximately 70–80% of the UK's nitrate input to the water environment comes from diffuse sources, with agricultural land as the main source. High levels of nitrate in drinking water can cause methaemoglobinaemia (blue-baby syndrome) in bottle-fed infants. The illness is caused by nitrite produced from nitrate in the gastro-intestine and is extremely rare, no cases have been recorded in the UK since 1972, but concern about high levels of nitrate in water has lead to a limit of 50 mg l^{-1} for nitrate in public water supplies (Chapter 1). Whilst compliance for nitrate is high in England and Wales (>99% in 2003) a total of 94 nitrate removal schemes are proposed for the period from 2005–2010.

Conventional drinking water treatment does not remove nitrate, and specific treatment processes are required for removing nitrate to acceptable levels. The simplest solution is to blend high nitrate water with low nitrate water. This is a relatively low-cost option but is dependent on a source of low nitrate water being available for blending. However, it is often not viable as either low nitrate water is not available or because of the expense of transferring low nitrate water over long distances. Removal of nitrate is therefore required and a number of processes have been developed including ion exchange, reverse osmosis, electrodialysis and biological denitrification.

11.2.1   Ion exchange

Ion exchange resins exploit functional groups that are initially bonded to chloride ions. When nitrate-rich water flows over the resin beads, the chloride ion is exchanged for a nitrate ion because of its relatively higher affinity for the quaternary amine group (see summary below). The chloride ion flows out with the treated water and the nitrate ion remains bonded to the functional group. When all of the resin's functional groups have been bonded to contaminant anions, the resin is saturated. The resin is then regenerated with 2–3 bed volumes of a saturated brine solution.

Functional group: strong base quaternary ammonium
Exhaustion: $RCl + NaNO_3 \Rightarrow RNO_3 + NaCl$
Regeneration: $RNO_3 + NaCl \Rightarrow 2RCl + NaNO_3$

The resin also attracts similar anions including carbonate and sulphate, most anion exchange resins have a higher selectivity for sulphate than nitrate, so nitrate specific resins have been developed. The regeneration results in a waste brine solution high in nitrate, sodium, chloride, sulphate and other ions. Dealing with the disposal of regenerant waste is a major concern in the operation of ion exchange facilities and disposal may be an issue in NVZs. Recently there has been interest in the use of electrochemical cells to reduce the nitrate in the waste brine to nitrogen gas.

An ion exchange plant normally consists of two or more resin beds contained in pressure shells with appropriate pumps, pipework and ancillary equipment for regeneration. The pressure shells are typically up to 4 m in diameter and contain 0.6–1.5 m depth of resin.

11.2.2 Reverse osmosis

In the reverse osmosis (RO) process, pressure is applied to water to force it through a semi-permeable membrane leaving the majority of impurities behind. Typically 85–95% removal of nitrate can be achieved although this will depend on many factors such as the initial quality of the water, the system pressure, type of membrane and water temperature. RO is used to treat a proportion of a flow which is then blended back into the main flow to produce a water of the desired nitrate concentration (Figure 11.1). As with the ion exchange processes the RO process produces a high-nitrate brine solution that requires disposal.

11.2.3 Electrodialysis

Electrodialysis is an electrochemical process in which ions migrate through an ion-selective membrane as a result of their attraction to the electrically charged membrane surface. A positive electrode (cathode) and a negative electrode (anode) are used to charge the membrane surfaces and to separate contaminant molecules into ions. The process relies on the fact

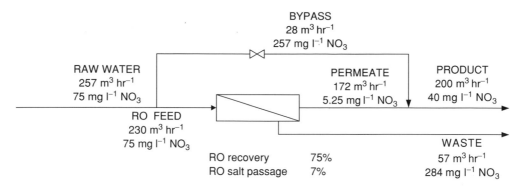

Fig. 11.1   Example of an RO plant design to produce 200 m³ hr⁻¹ of water containing 40 mg l⁻¹ of nitrate from a 75 mg l⁻¹ supply.

that electrical charges are attracted to opposite poles, for example it will separate nitrate from cations such as calcium and magnesium. A single-stage electrodialysis system will remove ~50% of TDS in water, allowing water to be blended to achieve the 50 mg l⁻¹ nitrate target.

11.2.4   Biological denitrification

Many of the chemical nitrate removal methods do not transform nitrate to benign products but generate brine waste liquors. An alternative approach is to use biological denitrification. Here heterotrophic bacteria, under anaerobic conditions, use nitrate as an electron acceptor for electrons and hydrogen and allows the nitrate to be reduced to the gaseous forms of nitrogen (N). This process occurs naturally in groundwater under anaerobic conditions when there is a supply of organic carbon adequate to sustain denitrification activity. Denitrification can be summarised by

$$2NO_3 + C_2H_5OH \rightarrow N_2 + 3H_2O + 2CO_2$$

One example of where biological denitrification is used is in the in-situ denitrification of groundwaters. Here treatment depends on producing favourable conditions for the development of heterotrophic populations of denitrifying bacteria. This is achieved through the injection of an organic carbon source (organic compounds tested include methanol, ethanol, glucose, acetate, formic acid) being utilised during the process into the aquifer, which creates a natural biologically anoxic active zone (Zone I in Figure 11.2) in the vicinity of the well where nitrate can be reduced to N gases. Zone II serves as a sand filter in which the turbidity and suspended solids are removed. Zone III contains lower concentrations of nitrate.

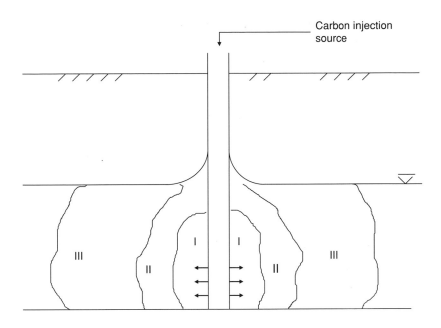

Zone I - Biochemical reactor
Zone II - Filter
Zone III - Storage of low NO_3^- water

Fig. 11.2 Schematic description of in situ denitrification (adapted from Cartmell, 1997).

11.3 Bromate

Bromate (BrO_3^-) is classified by the International Agency for Research on Cancer (IARC), a part of the WHO, as a Group 2B or 'possible human' carcinogen and EU directive 98/83/EC (Drinking Water Directive), adopted in November 1998, specifies a maximum bromate concentration of 25 $\mu g\,l^{-1}$ by 2003 and 10 $\mu g\,l^{-1}$ by 2008. The US Environmental Protection Agency Stage 1 Disinfectants/Disinfection By-Products rule, also signed in 1998, set a maximum bromate level of 10 $\mu g\,l^{-1}$. In the UK, the Water Supply (Water Quality) Regulations 2000 specified a maximum bromate level within drinking water of 10 $\mu g\,l^{-1}$.

Contamination by bromate is commonly associated with disinfection by-product formation during the ozonation of water containing bromide (Br^-). Bromide is not regulated and is found naturally within most water systems with concentrations in freshwater typically around 15–200 $\mu g\,l^{-1}$. Slightly higher levels are found in groundwater regions with saltwater intrusion. The mechanism of bromate formation is complex and can be by direct oxidation by molecular ozone (O_3) or indirect oxidation by the hydroxyl radical (OH). Based on chemical stoichiometry 1 μg of bromide can form 1.6 μg of bromate but in practice conversion is only about 30%.

$$Br^- \xrightarrow{\quad O_3 \quad} BrO^- \xrightarrow{\quad O_3 \quad} BrO_2^- \xrightarrow{\quad O_3 \quad} BrO_3^- \quad \text{Direct mechanism}$$

Bromate formation is very dependent on water quality and ozonation conditions and the best way to achieve standards is to control its formation; pH has the greatest impact on bromate formation. A number of studies have shown that decreasing the pH from 8 to 6 will reduce bromate formation by over 50% (von Gunten, 2003). Ammonia (NH_3) addition has also been suggested as a control method as it does not affect the ozone oxidation and disinfection processes but reacts with HOBr, an intermediate in the indirect oxidation of bromide to bromate by hydroxyl radicals. Once bromate is formed, as with nitrate it will not be removed easily by conventional water treatment processes. A wide range of processes have been investigated including membranes, activated carbon, UV or the use of reducing agents such as zero-valent iron.

11.3.1 Membranes

Membrane processes are able to separate anions from water, and nanofiltration has been reported to achieve bromate removal of between 75–100%. As with nitrate a concentrated waste stream is produced, which would need remediation prior to disposal. Finally, the cost for bromate treatment alone is high, and it is unlikely that membrane filtration would be cost-effective without significant process integration.

11.3.2 UV processes

Ultraviolet (UV) irradiation (wavelength 100–400 nm) is widely used for water disinfection but more powerful doses ($\sim$700 mJ cm^{-2}) have been shown to reduce bromate to bromite (BrO_2^-) and subsequently bromide. 20% reduction can be achieved at this dose and performance will be dependent on UV wavelength and power. Photocatalysis using titanium dioxide (TiO_2) catalyst can improve removal, but long reaction times mean that significant improvements in the process is required before it can be competitive.

11.3.3 Reducing agents

Reducing agents include sulphur compounds such as thiosulphate, sulphite and sulphur dioxide and many have been tested for removing bromate. Sulphite addition can reduce bromate levels by 60% although this needs high pH. Ferrous iron (Fe^{2+}) is also a promising reducing agent as it reduces bromate to bromide and oxidises into the ferric form (Fe^{3+}), with any remaining Fe^{2+} oxidised by dissolved ozone or oxygen.

$$BrO_3^- + 6Fe^{2+} + 6H^+ \Leftrightarrow Br^- + 6Fe^{3+} + 3H_2O$$
$$4Fe^{2+} + O_2 + 4H^+ \Leftrightarrow 4Fe^{3+} + 2H_2O$$

Studies using natural water showed bromate reduction occurs within 10 minutes, and equilibrium was reached after 15 min. Reduction is also pH dependent but here the higher rates are achieved at lower pH and dissolved oxygen (DO) levels. Increased pH and DO will lead to oxidation of Fe^{2+} to Fe^{3+}. Zero-valent iron (Fe^0) was first proposed as a remediation methodology for nitrate in the early 1990s. Bromate is reduced at a faster rate than either nitrate or chlorate by Fe^0, and 90% removals are achieved at an empty-bed contact time of 20 min during column trials, compared with only 10% for nitrate (Westerhoff *et al.*, 2003).

11.4 Arsenic

Soluble arsenic generally exists in either the +3 or +5 valence state, depending on local oxidation–reduction conditions. Arsenic (III) is also known as arsenite; arsenic (V) is also known as arsenate. Under anaerobic conditions, arsenic exists primarily as arsenic (III) (arsenite). Under aerobic conditions, arsenic exists primarily as arsenate (V). Arsenic is introduced into the aquatic environment from both natural and man-made sources. Typically, however, arsenic occurrence in water is caused by the weathering and dissolution of arsenic-bearing rocks, minerals and ores. Although arsenic exists in both organic and inorganic forms, the inorganic forms are more prevalent in water and are considered more toxic. The form of both arsenite and arsenate found in water is pH dependent, and over the pH range of natural water (pH 6–9) arsenite occurs as an uncharged species H_3AsO_3, whilst arsenate occurs as a charged species, $H_2AsO_4^-$ or $HAsO_4^{2-}$. This plays a major part in their treatment and because of the negative charge on the arsenate species they are easier to remove than arsenite in processes such as adsorption, ion exchange and coagulation (USEPA, 2000).

Table 11.2 Comparison of arsenic removal processes (USEPA, 2003).

Process	% Removal
Coagulation and filtration	95
Coagulation assisted microfiltration	90
Lime softening (pH >10.5)	90
Ion exchange (sulphate <50 mg l^{-1})	95
Activated alumina	95
Granular ferric hydroxide	98
Reverse osmosis	>95

A comparison of the performance of treatment processes is shown in Table 11.2 and the optimum pH range for the main arsenic removal processes is shown in Figure 11.3.

11.4.1   Coagulation

Coagulation is a widely used water treatment process and is effective for arsenic removal. Ninety per cent removal of arsenate is possible when

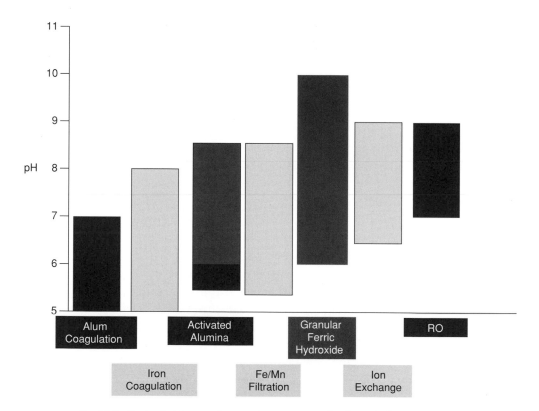

Fig. 11.3 Optimal pH ranges for arsenic treatment technologies (adapted from USEPA, 2003).

using ferric-based coagulants. Removal of arsenite is much lower and it needs to be oxidised to arsenate using chlorine, permanganate or ozone. Coagulation with ferric coagulants works best at pH below 8 whilst alum works best over the narrower range, from pH 6–7.

11.4.2 Ion exchange

Ion exchange resins are good at removing arsenate but not arsenite. As with the nitrate removal resins there will be competition from other ions such as sulphate. Typical conditions are EBCT of 1.5–3 min and if the sulphate levels are under 25 mg l^{-1} then a bed can typically treat several hundred to a thousand bed volumes before the resin must be regenerated.

11.4.3 Adsorbents

Adsorption is considered to be the best available treatment for arsenic removal from groundwaters. It is a simple process, has a reasonable bed life, produces minimal process residuals and the arsenic is strongly bound to the media, allowing safe disposal. A wide range of inorganic and organic adsorbents have been evaluated for arsenic removal of which the leading two are activated alumina (AA) and granular ferric hydroxide (GFH). A comparison of the two media is shown in Figure 11.4 where at natural pH 190 000 bed volumes were treated with GFH before the outlet concentration reached 10 μg l^{-1}, over 15 times greater than AA under the same conditions (Selvin *et al.*, 2002). Severn Trent Water developed the SORB 33^{TM} system which uses a media developed by Bayer called Bayoxide® E33 media. This process has been installed at a number of groundwater

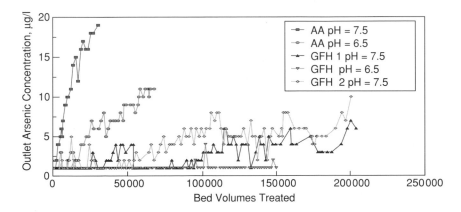

Fig. 11.4 Arsenic removal using ferric and aluminium media (source: Selvin *et al.*, 2002).

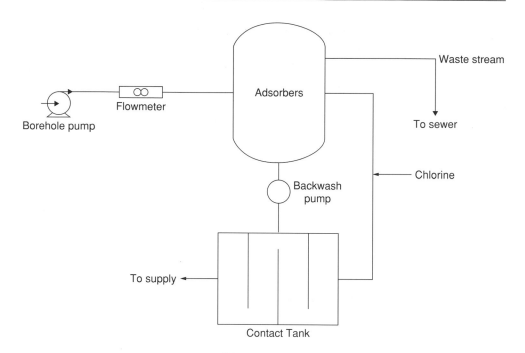

Fig. 11.5 Flowsheet of a SORB 33^{TM} arsenic treatment process (adapted from Selvin *et al.*, 2002).

sites in the UK. The SORB 33^{TM} is a simple process where pressurised water flows through a fixed-bed pressure vessel containing the Bayoxide® E33 media (Figure 11.5). As the contaminated water passes through the media, arsenic is adsorbed and removed to a level below the drinking water standard of 10 μg l^{-1}.

11.5 Iron and manganese

Iron and manganese occur naturally in borehole water in the reduced forms of Fe(II) and Mn(II) and are also found complexed in waters with high concentrations of organic matter. They are not normally considered to be a health concern but particular problems noted for iron and manganese include taste, colour and odour. They are both oxidised to some extent by exposure to air, producing oxide precipitates which can precipitate and stain surfaces and clothing. The reactions are shown below:

$$4Fe^{2+} + O_2 + 8OH^- + 2H_2O \Leftrightarrow 4Fe(OH)_3$$
$$2Mn^{2+} + O_2 + 4OH^- \Leftrightarrow 2MnO_2 + 2H_2O$$

Both iron and manganese can be removed from water by oxidising them to the insoluble forms, ferric hydroxide and manganese oxide, respectively.

Whilst this reaction is thermodynamically favoured it happens very slowly, particularly the oxidation of the manganous ion. Therefore, it is common to use an oxidant to speed up the reaction; common oxidants are chlorine, permanganate or chlorine dioxide. Figure 2.1 shows a water treatment flowsheet designed to remove both iron and manganese. Here the removal is achieved by changing the pH after flotation to 6.5 to remove iron (and aluminium), and then chlorine is added and the pH increased to ~pH 9 to remove manganese. Manganese removal can actually be autocatalytic, in that it is promoted by the presence of the manganic dioxide (MnO_2) product. However, in practice the amount of oxidant required is always less than that dictated by the above equation, leading to the postulation that Mn(III) oxide is formed by hydrolysis:

$$Mn^{2+} + MnO_{2(s)} + H_2O \Leftrightarrow Mn_2O_{3(s)} + 2H^+$$

Permanganate dosing coupled with a greensand filter – a sand filter laced with MnO_2 – can promote oxidation whilst allowing excess permanganate to be temporarily adsorbed on the filter.

11.6 Fluoride

The intake of excess fluoride can cause fluorosis which affects the teeth and bones. Moderate amounts lead to dental effects, but long-term ingestion of large amounts can lead to potentially severe skeletal problems. Low levels of fluoride intake can help to prevent dental problems and the current PCV for fluoride is 1.5 mg l^{-1}. Fluoride (F^-) is found in the environment as fluorides in various minerals including: fluorspar (mineral fluorite, CaF_2), cryolite (Na_3AlF_6) and fluorapatite ($Ca_5(PO_4)_3F$). The concentrations of fluoride in water are often limited by the solubility of fluorite but in the absence of calcium, concentration can be as high as 20 mg l^{-1}, however more typically rivers, lakes and groundwater have fluoride concentrations of less than 0.5 mg l^{-1}. Table 11.3 compares treatment processes for fluoride removal, and the use of activated alumina or reverse osmosis are considered the best available techniques.

Table 11.3 Comparison of fluoride removal processes.

Process	%Removal
Coagulation/filtration	5
Ion exchange (sulphate <50 mg l^{-1})	1–2
Activated alumina	85–90
Granular ferric hydroxide	1
Reverse osmosis	40–60

11.7   Lead

Lead does not normally occur in natural water supplies but gets into drinking water primarily from household plumbing systems containing lead in pipes, solder, fittings or the service connections to homes. Lead pipe was still in common use in the UK until the 1970s and it is estimated that nine million homes are still using it. The amount of lead dissolved from the plumbing system depends on several factors, including pH, temperature, water hardness and contact time. Low alkalinity and low pH waters are likely to cause the largest release of lead. The EU standard for lead is 25 μg l^{-1} and a final standard of 10 μg l^{-1} is to be achieved by 25 December 2013.

The most effective process for controlling lead is phosphate dosing which forms an insoluble layer of lead phosphate over the pipe walls. Phosphate can be dosed in two main forms, either as orthophosphoric acid or as one of the sodium phosphates, usually a solution of monosodium phosphate. Typically it would take a year for a full protective scale to build up, based on an initial dose of 1 mg l^{-1} of phosphorus, although temperature and the condition, age and history of the pipework being coated would affect this. The optimum conditions recommended by the DWI for orthophosphate concentrations and pH are:

- Orthophosphate residual – 0.7–1.7 mg P l^{-1}
- pH value – 7.2–7.8 for hard waters
- pH value – for soft waters a higher pH value than this range may be required depending on the organic colour and the need to minimise corrosion of iron distribution systems.

References

Cartmell, E. (1997) Aquifer denitrification: An experimental and modelling evaluation. PhD Thesis. Imperial College, London, UK.

Selvin, N., Upton, J., Sims, J. & Barnes, J. (2002) Arsenic treatment technology for groundwaters. *Water Supply*, **2**, 11–16.

USEPA (2000) Technologies and costs for removal of arsenic from drinking water. EPA 815-R-00-028.

USEPA (2003) Arsenic treatment technology design manual for small systems. EPA 816-R-03-014.

von Gunten, U. (2003) Ozonation of drinking water: Part II. Disinfection and by-product formation in presence of bromide, iodide or chlorine. *Water Res.*, **37**, 1469—1487.

Westerhoff, P. (2003) Reduction of nitrate, bromate, and chlorate by zero-valent iron (Fe$_0$), *J. Environ. Eng.*, **129**, 10.

Sludge Treatment and Disposal $\mathbf{12}$

12.1 Introduction

The objective of most water treatment processes is to remove contaminants from the water to make it safe to drink. The materials removed during the treatment process are often referred to as residuals and are primarily solids and carrier water. The purpose of sludge treatment is to receive and concentrate these residuals and dispose of them. The treatment, transportation and disposal of the sludge makes up a major fraction of the total water treatment costs and this chapter will look at the amounts and nature of sludge generated, as well as treatment, disposal and re-use options.

12.2 Sludge characterisation

Solids generated during the water treatment process consist primarily of nearly all the suspended solids (turbidity) in the influent, algae, viruses and dissolved solids as well as all the chemical added that form precipitates including lime, iron- and aluminium-based coagulants and chemicals such as PAC added to remove organics or to weight sludge blankets. A knowledge of the properties and quantities of sludge produced is essential for the design of any process flowsheet. A list of likely residues is shown in Table 12.1 and the quantity and nature of waterworks sludge is dependent not only on the treatment processes but also on the coagulant used (see Tables 12.2 and 12.3).

Although measurements of sludge flows and concentrations are difficult to undertake, the quantity of sludge produced can be calculated using solids balances based on raw water quality and chemical usage. Although most texts will give a method for calculating solids generation, an estimate

Table 12.1 Sludge production from a range of water types.

Source water	Average range of residue (kg dry solids per Ml)
Good quality reservoir	150–200
Fair quality reservoir	200–300
Average river water	350–450
Poor quality reservoir	400–500
Poor river water	500–600

can be made using the following guideline:

$$\text{Sludge solids (mg } l^{-1} \text{ treated water)} = 2 \times \text{turbidity (NTU)}$$
$$+0.2 \times \text{colour removed (Hazen)}$$
$$+2.9 \times \text{aluminium precipitated (mg } l^{-1} \text{ Al)}$$
$$+1.9 \times \text{iron precipitated (mg } l^{-1} \text{ Fe)}.$$

A 1998 survey of UK water companies was undertaken to gather data on sludge production and disposal routes (UKWIR, 1999). It identified that sludge production in the UK was approximately 131 000 tonnes ds (dry solids) per year. Alum sludges accounted (Figure 12.1) for the largest proportion (approximately 44%) followed by iron sludges (33%), softening sludges (17%), natural sludges (5%) and other WTW sludges (<1%).

When compared to wastewater sludges little interest has been paid to the characterisation and treatment of potable sludges. Potable sludge is made up primarily of coagulant flocs and can contain up to 20% coagulant metal ion. In between the flocs there is what is termed free water which is easily removed by mechanical methods but water bound on the flocs themselves (surface water) and the water held between colloidal particles (interstitial water) is not that easy to remove (Figure 12.2). The metal content of potable sludge has a significant effect on the bound water content and ferric sludges are typically easier to dewater than alum sludges. This phenomenon was initially attributed to higher density of iron flocs but it has been shown to be due to lower bound water content. Sludge storage time also has an effect on dewaterability. Organics also play a major role in the structure of sludges and the breakdown of organic material, for

Table 12.2 Sludge produced by conventional water treatment processes.

Sludge source	Solids concentration (%)
Upflow clarifier	2–5
Filter backwash water	0.1 (1000 mg l^{-1})
Flotation clarifier	2–5
High rate clarification	1–2

Table 12.3 Chemical nature of ferric and aluminium sludges (adapted from Godbold *et al.*, 2003).

	Ferric sludge	Alum sludge
Total solids (% of TS)	1.85–17.6	0.1–27
Volatile solids (% of TS)	–	10–35
Suspended solids (% of TS)	–	75–99
pH	–	5.5–7.5
COD (g l^{-1})	9–70	0.5–27
Aluminium (% of TS)	4.5–10	4–11
Iron (% of TS)	19–38	6.5
Phosphate (mg P l^{-1})	0.34–6.22	0.3–300
Arsenic (mg As l^{-1})	0.002	0.04
Total plate counts	–	30–30 000

example algae during sludge storage is known to inhibit dewaterability. The presence of natural organic matter absorbed onto coagulant flocs has also been shown to have a detrimental effect on sludge rheology, leading to a sludge that is harder to dewater.

12.3   Sludge treatment

An example flowsheet for sludge collection, treatment and disposal is shown in Figure 12.3, and includes sludge collection from a clarification process followed by gravity thickening, dewatering and drying prior to

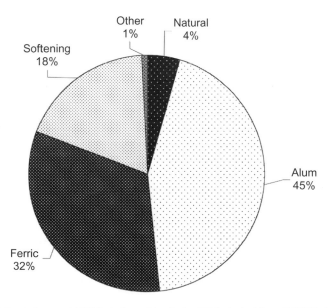

Fig. 12.1 Estimated annual WTW sludge arising in the UK (adapted from UKWIR, 1999).

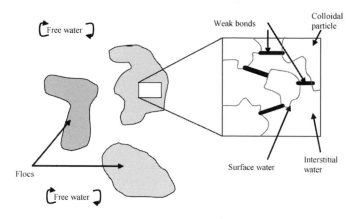

Fig. 12.2 Structure of coagulant sludges.

disposal. Table 12.4 compares the performance of different dewatering processes. The gravity thickening process can either be batch or continuous of which the WRc design of continuous thickener is a popular choice (Figure 12.4). This process has a flat bottom over which a rake with spiral blades draws settled sludge continuously towards the central outlet; this rake has a ploughing action that produces a thicker sludge than batch thickening (5–20%) and the supernatant from the thickener can be returned to the works inlet or discharged and is normally in the range 4–8 NTU.

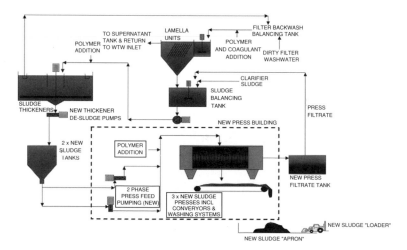

Fig. 12.3 Example of sludge treatment and disposal flowsheet.

Table 12.4 Typical sludge dry solids (%ds) produced during sludge dewatering.

Sludge treatment	Sludge produced (% ds)
Gravity thickener	5
WRc thickener	5–20
Superthickener	20
Membrane thickener	10
Centrifuge	10–20
Filter press	10–20
Belt press	10–15
Storage lagoon	10–15
Freeze thaw	40–80
Windrows	50

Once thickened, the sludge can be passed on to a dewatering stage such as a filter press or a centrifuge. Filter presses are one of the oldest methods for filtration and typically comprise a stack of filter plates held tightly closed by hydraulic pressure (Figure 12.5). The filter plates have either a machined or moulded filtration drainage surface that supports a filter media, typically a polypropylene filter cloth. A liquid/solid slurry is pumped into the chambers under pressure. Clear filtered liquid passes through the filter cloth, against the drainage surface of the plates, and out of the discharge ports. Solids are retained on the filter cloth forming a filter cake. The filter plates are separated and the filter cake is discharged and the process can achieve a solids concentration in excess of 20%. Gravity belt thickeners operate by allowing the water to drain from the sludge through porous belts. Performance is flexible depending on belt speed and power and polymer consumption are low. The filtrate is of good quality and the process is easily automated. Centrifugation can be applied to both thickening and dewatering and performs two functions: clarification and thickening with sludge particles being settled under centrifugal forces. The most widely used type of centrifuge in sludge dewatering is the solid bowl scroll type, also called a decanter centrifuge.

A helical screw spins at a slightly different speed, moving the accumulated sludge towards the tapered end, where further solids concentration occurs. The sludge is then discharged. An increase in the bowl length can increase the settling area and the retention time. This in effect will increase clarification. The pond depth can also be manipulated. A reduced pond depth followed by more beach will result in a drier cake. The bowl speed can be altered to increase the G forces, and hence improved drying of the sludge is achieved. By manipulating the bowl speed, a compromise can be found to achieve the optimum clarification, throughput, cake dryness and

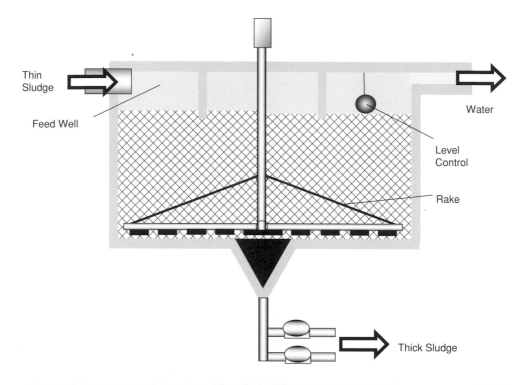

Fig. 12.4 Schematic and an internal view of a continous sludge thickener.

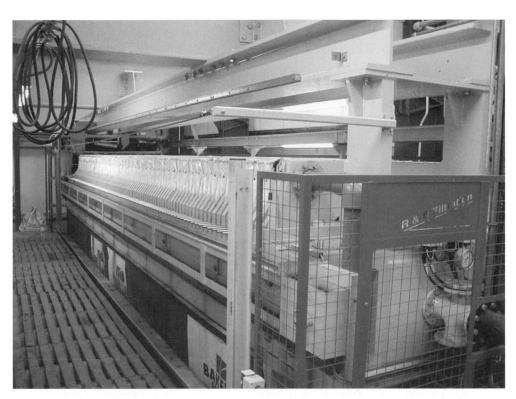

Fig. 12.5   Membrane filter press and an example of a filter press membrane plate (images courtesy of Yorkshire Water and South West Water).

maintenance costs. Any adjustments to scroll speed will affect the centrate and cake dryness. A lower differential results in a dryer cake, while a higher differential forms a wetter cake. The centrifuge is capable of producing cake at about 20% dry solids and has the advantage of being an automatic process. However, maintenance and power costs can be substantial and therefore, this process is usually found in large plants, where space is limited and skilled operators are available. Sludge drying in windrows is a final stage prior to disposal to landfill and can achieve around 50% dry solids.

Water treatment sludge is one of the most difficult sludges to dewater as up to 40% of the water is chemically bound to the particles. Conventional dewatering processes such as belt presses or centrifuges may be able to increase the total solids content up to 20%, but this is not usually the case. One process that can produce greater total solids is freeze–thaw conditioning followed by gravity filtration. The process of freeze–thaw conditioning changes the sludge from a suspension of small particles to a granular material that resembles coffee grounds (Martel & Diener, 1991). The mechanism responsible for converting sludge into granular particles is by ice crystal formation. Ice crystals are made solely from water therefore all other substances including floc particles are pushed away from the growing ice crystal. This forces the floc particles to become concentrated at the boundaries between ice crystals such that once freezing is complete, the sludge is no longer a suspension of fine floc particles but a matrix of ice crystals and consolidated floc particles or grains. When the ice crystals thaw, the grains remain consolidated and do not redissolve. These grains are large enough to settle easily under gravity in the clear meltwater. The sludge is dewatered by decanting or draining the meltwater. In cold climates, freeze–thaw conditioning is easily accomplished during the winter months in outdoor freezing beds. However, this technology is not usable at water treatment plants located in warm climates or at plants without the available land area. For these situations, a mechanical freezing device is needed such as the one shown in Figure 12.6.

12.4   Sludge disposal

The majority of WTW sludge is disposed to landfills (approximately 57%), with the disposal to wastewater treatment works (WwTW) being the next most common route (Figure 12.7). The cost per tonne dry solids of disposal of WTW sludges ranges widely between companies, depending on their chosen waste management technique. The mean cost is about £41 per tonne dry solids and the highest reported cost, £323. A number

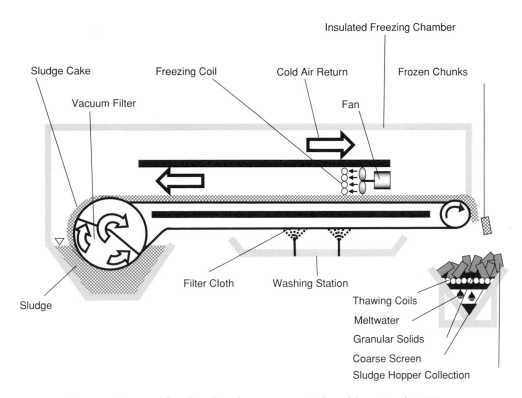

Fig. 12.6   Proposed flowsheet for a freeze separator (adapted from Martel, 1996).

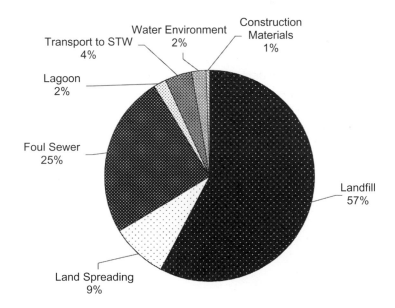

Fig. 12.7   Reported disposal routes for WTW sludges in the UK (adapted from UKWIR, 1999).

of novel disposal routes have been identified including incorporation into construction materials such as bricks, concrete and synthetic aggregates (Godbold *et al.*, 2003).

One alternative disposal route for potable sludge is to recover the metal ion and reuse it as a coagulant. This idea is not new, it has only recently been considered as a truly economical and viable process. The first attempt at coagulant recovery was made by W. M. Jewell in 1903 who patented a process to recover aluminium from water treatment plant sludge by reaction with sulphuric acid. In 1923 W. R. Mathis patented an almost identical process whilst research work conducted in Britain, Japan, Poland and the United States in the 1950s raised the profile of the technique. Since then many pilot- and full-scale plants have been set up throughout the world using the process of coagulant recovery. Four methods for coagulant recovery have subsequently been developed. Acidification is by far the commonest; alkalisation, ion exchange and the use of composite membranes being the others. The amount of metal ion that can be recovered is dependent on the pH of recovery, sludge source and the dry solids content (Figure 12.8). There are two main objectives to all these coagulant recovery processes. The obvious one is the recovery of the metal coagulants for reuse in water or wastewater treatment, but removing the metal ion has the advantage of conditioning the sludge to minimise the total volume and mass for subsequent disposal.

Coagulants have primarily been used in water treatment plants, but now many wastewater discharges require tight control of phosphorus levels, and to achieve this many companies have adopted chemical phosphorus

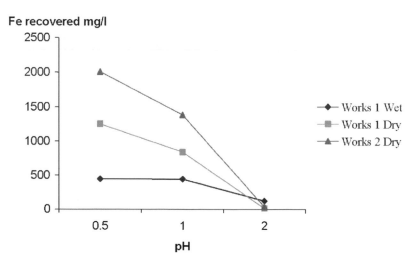

Fig. 12.8 Iron recovery from potable sludge collected from two lowland water treatment works.

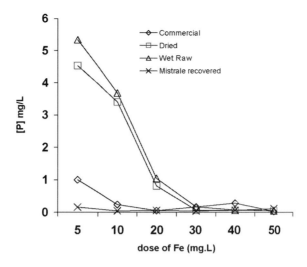

Fig. 12.9 Phosphorus removal from municipal wastewater comparing commercial and recovered iron coagulants.

removal using iron or aluminium coagulants. Iron coagulants recovered from potable sludge using acid (Figure 12.8) have been shown to be very effective at reducing phosphorus concentrations to well below the standard of 1 mg l^{-1} (Figure 12.9). One recent development is the Aqua Reci Process which is a coagulant recovery process based on Chematur Engineering's patented SCWO process (Aqua Critox®) and chemical methods developed by Feralco AB to recover metal salts. This oxide is separated from the liquid by filtration to DS of 20% and acidified with sulphuric or hydrochloric

Fig. 12.10 Manufactured control bricks (100% clay) on the left; trial bricks (10% alum sludge and incineration ash content) on the right (with kind permission of WRc).

acid. This is then separated again to produce a coagulant metal salt solution containing as much as 9% ferric or 3% aluminium depending on the feed sludge (Patterson *et al.*, 2003).

One of the most promising disposal routes for potable sludge is in the manufacture of bricks or cement (AwwaRF, 1990; Anderson & Skerratt, 2003; Godbold *et al.*, 2003). Godbold *et al.* (2003) reported the findings of a project that looked at reusing utility wastes including water treatment sludges. They successfully produced bricks made of 10% sewage sludge ash and aluminium sludge which were indistinguishable from 100% clay bricks (Figure 12.10).

References

American Water Works Research Foundation (1990) Slib, schlam, sludge. Cooperative Research Report with Keuringsinstituut voor Waterleidingartikelen, American Water Works Association, Denver, Colorado.

Anderson, M. & Skerratt, R.G. (2003) The inclusion of alum-based waterworks sludge (WTR) in commercial clay building bricks. *Tile & Brick Int.*, **19**(5): 328–333.

Godbold, P., Lewin, K., Graham, A. & Barker, P. (2003) Reuse of waste utility products as secondary commercial materials. Final Report, ENTRUST Project number 395069.018.

Martel, C.J., Affleck, R.T. & Yushak, M.L. (1996) A device for mechanical freeze–thaw conditioning of alum sludge, CRREL Report 96–15.

Martel, C.J. &. Diener, C.J. (1991) A pilot-scale study of alum sludge dewatering in a freezing bed. *J. Am. Water Works Assoc.*, **83**(12): 51–55.

Patterson, D.A., Jäfverström, S. & Stendahl, K. (2003) Recycling of coagulants from treated drinking water sludges. *CIWEM International Conference on the Management of Wastes from Drinking Water Treatment*, London, September 12–13.

UKWIR (1999) Recycling of water treatment works sludges (Ref: 99/SL/09/1). UK Water Industry Research Limited.

Index